THE VISUAL LIFE OF CLIMATE CHANGE

Saffron O'Neill

First published in Great Britain in 2026 by

Bristol University Press
University of Bristol
1-9 Old Park Hill
Bristol
BS2 8BB
UK
t: +44 (0)117 374 6645
e: bup-info@bristol.ac.uk

Details of international sales and distribution partners are available at bristoluniversitypress.co.uk

© Saffron O'Neill 2026

The digital PDF and ePub versions of this title are available open access and distributed under the terms of the Creative Commons Attribution-NonCommercial-NoDerivatives 4.0 International licence (https://creativecommons.org/licenses/by-nc-nd/4.0/) which permits reproduction and distribution for non-commercial use without further permission provided the original work is attributed.

DOI: 10.51952/9781529250046

British Library Cataloguing in Publication Data
A catalogue record for this book is available from the British Library

ISBN 978-1-5292-5002-2 paperback
ISBN 978-1-5292-5003-9 ePub
ISBN 978-1-5292-5004-6 OA PDF

The right of Saffron O'Neill to be identified as author of this work has been asserted by her in accordance with the Copyright, Designs and Patents Act 1988.

All rights reserved: no part of this publication may be reproduced, stored in a retrieval system, or transmitted in any form or by any means, electronic, mechanical, photocopying, recording, or otherwise without the prior permission of Bristol University Press.

Every reasonable effort has been made to obtain permission to reproduce copyrighted material. If, however, anyone knows of an oversight, please contact the publisher.

The statements and opinions contained within this publication are solely those of the author and not of the University of Bristol or Bristol University Press. The University of Bristol and Bristol University Press disclaim responsibility for any injury to persons or property resulting from any material published in this publication.

Bristol University Press works to counter discrimination on grounds of gender, race, disability, age and sexuality.

Cover design: Lyn Davies Design
Front cover image: Annie Spratt; Shawn; Marco Mons; Hyewon Hwang; NASA/UnsplashLaura Penwell; Santiago Manuel De la Colina; Yelena Odintsova; Marek Piwnicki; Ben Vloon/PexelsUN Climate Change, Kiara Worth/FlickrClimate Stripes, Ed Hawkins/University of Reading
Composition by Veronica White

To my children, for your love, resilience and energy.

To my PhD students, for your enthusiasm and insights.

I have learnt so much from you.

May you all continue to shape the world to be a more equitable place.

Contents

List of Figures		vi
Acknowledgements		viii
1	Introduction: 'Just Tell Me, What's the Best Climate Image?'	1
2	Adaptation: Heatwaves and Sea Level Rise	11
3	Impacts: Polar Bears and Flooding	29
4	Energy: Smokestacks and Wind Turbines	46
5	Science: Climate Stripes and Weather Maps	62
6	People: Politicians and Protesters	83
7	Conclusion: The Flow and Friction of Climate Images	100
References		125
Index		157

List of Figures

2.1	Typical 'fun in the sun' heatwave image	13
2.2	Typical 'idea of heat' image	14
2.3	Alternative heatwave imagery	16
2.4	Typical aerial view imagery to depict sea level rise threat to coral atolls	20
2.5	Typical sea level rise imagery, featuring a young Tuvaluan girl	21
2.6	The 'wishful sinking' media narrative, Tuvalu	23
2.7	Alternative visions of Tuvalu adapting to the impacts of sea level rise	27
3.1	Typical polar bear image	31
3.2	Cartoon parody of the CRU 'Climategate' email hacking	34
3.3	Protester dressed in a polar bear costume	35
3.4	Composite image of a polar bear on a small ice floe surrounded by sea	36
3.5	Typical flood rescue image	39
3.6	Photograph from David Maurice Smith's Lismore flooding series	40
4.1	Typical smokestacks image	48
4.2	Colour wheel of images arising from a Getty Images search for 'smokestacks'	50
4.3	Typical wind turbine climate change image	53
4.4	Colour wheel of images arising from a Getty Images search for 'wind turbines'	54
4.5	Typical image featuring a heat pump	57
4.6	An energy story created using a climate solutions journalism approach	59
5.1	The 'Reasons for Concern' (or 'Burning Embers') IPCC diagram	64
5.2	The crocheted Climate Stripes blanket, inspiration for the Warming Stripes	68
5.3	Reading Football Club display the Climate Stripes on their team shirt	70
5.4	The 'Climate Generations' figure	71

5.5	Purple began to feature at the top end of the Australian Bureau of Meteorology's weather map scale from 4 January 2013	76
5.6	'It's called summer' meme	78
6.1	Typical image featuring political figures	86
6.2	Colour wheel for images from COP26 (Glasgow) news coverage	87
6.3	Typical image arising from a UNFCCC COP	88
6.4	Typical image featuring a performance-led protest	90
6.5	'Condescending Wonka' meme	96

Acknowledgements

Thank you to my PhD students, present and past. Our conversations have shaped and inspired the work in this book in so many ways. Particular thanks to Sylvia Hayes on visual journalism and protest imagery; Ollie Blewett for insights on climate, mobility and decolonisation; Hannah Hayes on communication of flooding; and Ned Westwood on the role of memes in climate communication. An especially big thanks to Annayah Prosser, who read each chapter and made many valuable suggestions. Ex-PhD student, now communications consultant Niall McLoughlin developed several short-form videos for this book, including filming interviews with some of the people featured herein; this experience played an important role in helping me refine the book's storyline. Finally, thanks to Veronica White for not only her insights on photography, imagery and AI, but for nailing the brief in creating the book's cover image. Thanks, all, for your constructive critique and support; it has been very much appreciated.

A big thanks to my collaborators in the Centre for Climate Communication and Data Science (3CDS), especially Co-Director Travis Coan. Warm thanks to Ranu Malla, Ned Westwood and Tristan Cann for our conversations across the social-computational science boundary, which resulted in two of the figures in Chapter 4 (and the emergence of some exciting new work). Tristan also proofread the whole book.

This book was begun during my Leverhulme Research Fellowship. I am extremely grateful to the Leverhulme Trust for this funding. It allowed me to reboot my research (and stay in academia) after two pregnancies and maternity leaves, five years of part-time working and all the disruptions caused by being a 'sandwich generation' carer of disabled family members through pandemic lockdowns. More of these sorts of funding schemes are needed to support people to stay in academia who are on non-standard academic career paths (even if pregnancy, maternity leave and caring responsibilities are actually pretty standard life activities, at least outside of academia).

I should acknowledge the role James Painter and Mike Schafer, both scholars in climate communication research, played in this book's genesis. It was during a trip to Oxford University during Mike's sabbatical in 2019 that they both encouraged me to write this book, which until that point had been a much vaguer proposition. Some of the book's ideas were also worked through during a visit to the Visual Politics programme at the University of Queensland in 2023, part-funded through a University of Queensland–Exeter University (QUEX) grant. Special thanks to Roland Bleiker for his generous hosting and inspiring research.

An important source of knowledge, engagement and ground-truthing of my research has been the UK climate journalism organisation, Carbon Brief. My sincere thanks to all the team, led by Leo Hickman, for many productive conversations which have led in more or less direct ways to this book's content. Another input from beyond academia has been Alastair Johnstone, Climate Visuals programme lead at Climate Outreach. Alastair's input and insights (especially given his previous experience as a newspaper picture editor) have been invaluable. Thanks also to feedback from Katherine Dunn, Diego Arguedas Ortiz and the many Fellows of the 2022–2024 cohorts of the Oxford Climate Journalism Network (OCJN), part of the Reuters Institute for the Study of Journalism at the University of Oxford. Finally, I should express my gratitude to the many others working in the visual media industry and beyond (journalists, editors, photographers, campaigners and others) who have been generous in sharing their time and expertise over the last couple of decades, which have led in perhaps more indirect (but still important) ways to the content in this book.

Thanks to colleagues at the Tyndall Centre for Climate Change Research at UEA, a welcoming, open and interdisciplinary research centre – and where my interest in the visual communication of climate change began during my PhD studies. I hope this interdisciplinary book is true to that Tyndall Centre ethos.

This book has benefitted from discussions and encouragement from many Exeter Geography colleagues. Thanks to Neil Adger for comments and encouragement on the original proposal, and to Karen Bickerstaff and Cordy Freeman for keeping me on track in early days of our 'book club' meetings.

On specific chapters, I have many collaborators and colleagues to thank for preliminary discussions, for allowing me to reprint their images, and for reviewing draft chapters. In Chapter 2, I acknowledge scholar Carol Farbotko and scholar-activist Taukiei Kitara. Thank you for sharing your deep knowledge and connection to Tuvalu with me, and allowing me to share two photos. In Chapter 3, thanks to photographer David Maurice Smith, for allowing me to reprint his Lismore flood image and for generously discussing that work; and to Ella Buckland for allowing me to reprint the image of her and her daughter, Myla. In Chapter 4, thanks to Science and technology studies (STS) scholar Ola Michalec and environmental social scientist Patrick Devine-Wright for discussions around imaginaries of energy futures; and to Jodie Bond at the Local Storytelling Exchange, photographer Alex Leat and Rose Lewis for letting me use their picture. In Chapter 5, thanks to Ed Hawkins for sharing insights about the Climate Spiral and Climate Stripes images and to Reading FC for allowing the reprint of one of the team's pictures; to Andy Revkin and Andrew Goodwillie on interpretations of the Climate Stripes; to Helen Roberts and Aidan McGivern at the Met Office on weather and climate science communication; and to Arlene Birt for documenting the genesis of the climate generations figure. Many thanks to Julie Arblaster for insights on the now-famous purple weather map; and to Jen Catto and Duncan Ackersley for allowing me to share their photo of daughter Rose wrapped in the original climate stripes blanket. In Chapter 6, thanks to activist and artist Doug Francisco for insights on the Red Rebel Brigade.

Pitching my book to publishers ('if it has colour images, is it a "coffee table" book?') demonstrates that publishing scholarly work involving images can be much more complicated than research concerning words alone. Dealing with image copyright, licencing and permissions has been a minefield and made this a much longer project than anticipated. For this, I am grateful to the patience of the team at Bristol University Press. Thank you especially to Emily Watt and Anna Richardson, who saw that having colour pictures was a fundamental part of my original book proposal.

Finally, thank you to my family: from James' constant support, insights from climate science and barista skills (perhaps more than anything else, something that helped move the book forward!), to Rowan's skills in drawing, to Finn's complete fascination for the world viewed through a digital camera – you have shaped the thinking behind this book, and kept me happy and healthy.

1

Introduction: 'Just Tell Me, What's the Best Climate Image?'

This is a book about climate imagery: what climate change currently 'looks like', and what effects these images can have in the world. It is a book about how we should look more critically at images. It also provides a starting point for how we might go about making climate visuals more representative, equitable and just. But first, why should we care about images?

Think of a scientist. What image do you have in your head? Now, what about famine. Is there a picture that comes to mind? And finally, what visuals spring to mind when you imagine artificial intelligence (AI)?

Perhaps the scientist you've brought to mind has particular characteristics: male, white, able-bodied, middle-aged or elderly, wearing a white lab coat, possibly wearing spectacles or lab goggles. Maybe this scientist of yours even has crazy hair. These images suggest science means laboratory-based experimental work, carried out by people of a particular race, gender, age and physical ability. If your scientist had wild hair and their goggles were askew, perhaps they were also something of a 'mad scientist', carrying out secret experiments (and you may recognise the school kids in Chambers' drawing experiment who sketched signs on their scientists' lab walls such as 'keep out' and 'private'; Chambers 1983: 264). If the scientist in your mind has some of these features, it would be consistent with imaginations of a scientist which have remained consistent for 50 or so years or more, particularly in the Global North: scientists are male, white, do lab-based science only, and may work to challenge the mainstream (Mead and Metraux 1957, Chambers 1983, Moreau and Mendick 2012, Loverock and Hart 2018).

Feminist historian Laura Briggs writes powerfully about the way famine is visualised. She details how certain conventionalised images are conjured up when famine is imagined: an imploring, skeletal child holding an empty bowl; or a mother cradling her malnourished infant (a reimagined, secular Madonna and child). Common to these images is also what they lack: no visible home or shelter, no men, no supportive others; neither do they depict military roadblocks redirecting food supplies or debates in parliament (Briggs 2003). Briggs is clear that these images distract attention away from fundamental issues of power, structure and agency; and refocus them on a racialised ideology of the need for rescue. Geographer David Campbell writes about similarly racialised and gendered images emerging from the Darfur conflict to study not only images' content, but to also discuss how they emerge through photojournalism; that is, the fundamental question of these images is concerned with their power: what they do, how they function and their impact (Campbell 2010).

Both the famine and the scientist examples are examples of visual imaginings that have existed for many years. But what of issues which are still emerging in public discussion? Are visuals less constrained, more diverse? Perhaps you pictured AI as a white humanoid robot, which had visibly female features, against a blue background; or a humanoid arm tapping a keyboard, again with a blue background; or perhaps an outstretched human finger reaching out to touch the finger of a robot? If so, you are bringing to mind dominant visual tropes about AI. Far from being expansive and diverse, the visual discourse of AI is also narrow and problematic. Science Communication and AI ethics experts Kanta Dihal and Tania Duarte's (2023) report 'Better Images of AI' is clear that the images which dominate visualisations of AI do us a disservice. A burgeoning body of literature describes the racism and sexism inherent in AI, which contributes to these problematic visualisations (for example, Cave and Dihal 2020, Strengers and Kennedy 2020, Erscoi et al 2023). Further, when images show, for example, a disembodied robot hand tapping on a keyboard, they hide the accountability of humans who develop the technology, as well as confuse AI with robotics (anyway, why go to the lengths of inventing a disembodied robot arm to use a keyboard designed for a human, in order to perform an AI task?). Computer Scientist Stuart Russell's BBC Reith Lecture is enlightening, here, in discussing the role of AI in warfare; that if we visualise the role of AI as a Hollywood-style 'rampaging Terminator robot' we are not imagining, and importantly then regulating for, AI in future warfare – a reality that is actually already here, but looks nothing like the Terminator (Russell 2021).

These examples give some insight into why we should care about images. Images are not simply facile illustrations accompanying the 'serious' work of textual communication, but have substantial power to shape the world around us: Black women scientists face systemic racism and sexism, which the dominance of white men scientist imagery both reflects and perpetuates; racialised and gendered images of war and famine distract us from the root causes of conflict and their potential resolution; disembodied white robot arms and Terminator robots fail to illustrate the moral and ethical benefits and harms of AI.

These examples also show how images move and slip between different forms and categories. Images are not just a picture in a newspaper, on your Instagram feed, in a gallery, a shot in a video, or in a child's scribbled drawing stuck to the fridge door. As Davoudi and Machen (2022: 205) explain in their exploration of the terms 'image', 'imagination' and 'imaginaries', 'to imagine is to think about possibilities other than possible, times other than now, places other than here, and ideas other than made known' (see also Liao and Gendler 2019). The imagination is the domain of images, where imagery is more than visually-based mental representations, but also associated sights, sounds, ideas and words (Leiserowitz 2006, Davoudi and Machen 2022). While the imagination, then, is a faculty of an individual mind, an imaginary is a capacity of a political collective (Davoudi and Machen 2022).

So, then, to one of the foremost challenges of our time – climate change – and its connection to visual imagery and the imagination. While there is widespread public support for climate action globally (Andre et al 2024), action remains insufficient to address the challenge (UNFCCC 2023). Images play a fundamental role in how we imagine transitions and change; such as that needed to address climate change. Images shape and influence the futures we can imagine, desire and work towards. As these examples show, images are not simply facile illustrations accompanying the 'serious' work of communication delivered via other communicative modes such as text. Images have substantial power to shape the world around us, in terms of how we might respond to climate change. They are a fundamental way that we come to make sense of the world – both as depictions of what has happened, and imaginations of what might come to pass.

Geographer Gillian Rose discusses how digital technologies must force us to think beyond images as stable cultural objects and explore how images are produced, circulated and modified (Rose 2016). It can be helpful to think of four overlapping and interfacing sites in which meanings about images are made and shared: production, the image itself, circulation and audiencing (Rose 2023). In the case of the newspaper photo, for example, this encourages questions of understanding how technology and newsroom culture interface to produce a particular newspaper image, how this image travels and spreads through different online platforms as a digital object and perhaps how the image may be reused and reinterpreted in meme form or through its placement in an image stock library. At all points, the image is actively read and (re)interpreted. Images in this broad sense are a fundamental part of the performance of imaginaries, influencing the futures we may desire and work towards. In short, they are 'pivotal to the production of contemporary geopolitics' (Campbell 2010: 358).

Images are a fundamental part of communication. Images predate written text by thousands of years: humans were creating rock art by ~40,000 years ago as far apart as Europe and Indonesia (Aubert et al 2014), whereas written language emerged around 3,400 BC. They are a huge part of everyday life. A staggering 5 billion photos are taken every day now, for example, with the average person in the US taking 20 photos every day (Broz 2023). As International Relations scholar Roland Bleiker explains, the speed at which images circulate and the potential reach they may have (or 'go viral'; Bleiker 2018) is unprecedented. Imagery can be vibrant, colourful, memorable – and engage us emotionally (Joffee 2008). And, images can help us to remember, and to link related issues together (Graber 1990, Domke, Perlmutter and Spratt 2002). Yet despite these qualities, images are studied less than text across many disciplines, from Geography, to Media and Communications, to Politics and International Relations (Bleiker 2018, Aiello and Parry 2020, Rose 2023). This is also true of climate change communication, where, for example, studies of climate content in newspaper coverage often only analyse text, or look at TV coverage but only analyse written transcripts. Indeed, a meta-review of the literature on framing climate change did not include any studies of visual framing (Badullovich, Grant and Colvin 2020); while another meta-review of the environmental communication literature found only 2 per cent of studies analysed visual content (Takahashi et al 2021).

Framing for understanding visual climate communication

Images can be interpreted without training (unlike written language, which must be taught), and can cross linguistic boundaries (Popp and Mendelson 2010). However, this does not mean that images are always subject to the same reading or interpretation, as images are always 'read' or decoded within a particular socio-political context (Hall 1973); that is, readers need to share cultural references in order to interpret an image similarly. In Media and Communication Studies, scholars often work with the theoretical concept of 'framing'. Framing is an approach used throughout the social sciences, and has been criticised for lacking conceptual coherence and being operationalised in rather woolly ways – partly stemming from the concept arising from several different disciplines and traditions over a long period of time (see Schäfer and O'Neill 2017 for a review of frame analysis for climate communication research).

Two types of frames are relevant for this discussion. First, frames exist in communicating news media items, as media frames. Media framing is defined here following Political Communication scholar Robert Entman (1993: 52), that is, the 'selection of some aspects of a perceived reality to make them more salient … to promote a particular problem definition, causal interpretation, moral evaluation, and/or treatment recommendation' (Entman 1993: 52). Partly because of the imperative to communicate information within the limits of a media item (such as in a TV news broadcast, newspaper article or social media post) but also as a consequence of ideology, the structures and routines of the newsroom, journalistic norms and the influence of pressure groups and elites, some aspects are necessarily emphasised in a media item, while others are not (Entman 1993).

Second, media frames work as organising themes or ideas because they play to individual schemas which exist in people's minds. These culturally-specific mental classifications guide individual information processing (Scheufele and Nisbet 2007, Nisbet and Newman 2015). So then, these schemas in people's minds are called individual frames: 'mentally stored clusters of ideas that guide individuals' processing of information' (Entman 1993: 53). The applicability model (Price and Tewksbury 1997) posits that it is the interaction between media frames and individual schemas which influences a person's perceptions and attitudes. Media frames influence perceptions of an issue when they are applicable to an individual's existing interpretative schema: effects are seen where there is an overlap between a media frame and an individual's interpretative schema (Scheufele and Nisbet 2007; although note that frame incongruence – a mismatch between the two, can also have an effect on audiences). Likewise, individual frames (for example, of journalists) feed back into the media frames which come to exist (Engesser and Brüggeman 2015). So, both media and individual frames are highly interconnected (Entman, Matthes and Pellicano 2009).

Visuals are a key part of setting the frame. When certain types of visual frames come to dominate, others necessarily slip down the agenda – potentially even disappearing from sight. This can sideline particular perspectives, limit public debate, influence public opinion and can even affect regulatory options (Nisbet and Huge 2006). So, although images – and especially photographs – are often assumed to be portraying an objective reality (Urry 1992, Messaris and Abraham 2001) and taken at 'face value', the process of visual framing is deeply ideological (Hall 1973). Framing research has mostly concentrated on text (as mentioned previously, images are far less studied than text in the academy; Bleiker 2018, Aiello and Parry 2020, Rose 2023). The use of framing for understanding visual communication here looks to research by scholars in Media, Communication and Journalism; including Mary Bock, Erik Bucy, Renita Coleman, Shahira Fahmy, Betsi Grabe, Katy Parry and Wayne Wanta (see for example, Fahmy 2005, Fahmy, Kelly and Kim 2007, Grabe and Bucy 2009, Coleman 2010, Parry 2010, 2011, Fahmy, Bock and Wanta 2014).

With this introduction to visual communication in mind, the discussion now moves onto the particularities of climate change visuals. It starts by considering one of the most common enquiries I get about my research: 'Just tell me, what's the best climate image?'

Climate change has an image problem

What is the best climate image? It might seem like a straightforward question. But the short answer is – as, you might have guessed, given the literature just reviewed – that there is no 'best' image. What constitutes 'best'? For whom, in which circumstances, and whose voices are privileged (and marginalised) in using this visual? The longer answer is that climate change has something of an image problem.

The greenhouse gases – carbon dioxide, methane, nitrous oxide and others – that contribute to a warming planet are colourless and invisible. In addition, the definition of 'climate change' inherently refers to a process happening over a long period, rather than a snapshot, of time (it takes 30 years to define an 'average' climate, according to the World Meteorological Office definition, WMO 2024). These factors mean that climate change, as a fundamental concept, cannot be 'seen'. These factors help to explain the problematic status of visuals for representing climate change (Doyle 2009). But this does not mean that there is no such thing as a climate visual. On the contrary, as climate change has become one of the foremost challenges of our time, this complex concept has been increasingly represented via photographs, drawings, paintings, films, scientific figures, adverts, cartoons and memes, among many other types of visual renderings. But, climate imagery remains largely confined to a narrow set of visual tropes: such as polar bears, melting icebergs, wind turbines, the Earth in space or a politician behind a lectern. Climate images are often distant from people's everyday lives and experience. This is true both for the imagery in our own minds (or 'affective imagery'; see Leiserowtiz 2006; Leviston, Price and Bishop 2014) and the imagery that we see in the world around us, for example on TV (León and Erviti 2013), in newspapers (O'Neill 2013) and online (Pearce et al 2018).

Recognising the narrow bounds of this visual imaginary of climate change is important, and brings three broad challenges. First, it distances people from the issue. This can have the effect of displacing and marginalising the vulnerability of ecosystems, biodiversity and human communities to climate change. It can also disconnect people from the fundamental issue at the root of climate change, that is, the need to move away from fossil fuels (in environmental coverage and campaigning, pristine nature imagery has long been critiqued for how it can exclude other possible representations, which decontextualise and disconnect the environment from the human actions causing its decline; Hansen and Machin 2013, Takach 2013). For example, as Chapter 2 explores, when heatwaves are portrayed as 'fun in the sun' through pictures of people having fun splashing about at the beach, the very substantial health dangers of heat extremes are ignored (O'Neill et al 2023, WHO 2024). In terms of images of climate and energy, geographical landmarks and people are often entirely absent from the climate-energy picture. As the research shows in Chapter 4, wind turbines and smokestacks imagery often fails to depict recognisable landmarks or landscapes, or show humans at all; perpetuating a placeless, technocentric approach to imagining energy futures. This echoes work exploring visual representations of climate change in search results of Google Images, which also showed a distinct lack of people in images in search returns (Pearce and De Gaetano 2021). Climate imagery which distances people from the issue can have the effect of making people feel like they are not able to take action, or that the action they can take is insignificant. The images of politicians featuring in Chapter 6 may dominate climate news, but tend to make people feel both that climate change is unimportant and that they feel disempowered to take action (O'Neill 2013, O'Neill et al 2013).

Second, current visual portrayals of climate change tend to exclude opportunities for imagining a more equitable and resilient future. In the heatwave example, this might be images of a city plaza planted up with new trees to increase shade and decrease the urban heat island effect. However, very few images currently portray how people and places might adapt to living in a world increasingly subject to heat extremes. Similarly in the flood images discussed in Chapter 3, there is a focus (especially in news visuals) on people in distress experiencing the immediate impacts of the flood event itself, but scant attention to visualise the ongoing economic and mental health impacts of flooding, or how adaptation to flooding might be imagined. The narrow bounds of climate visuals reflect and shape our lack of imagination, and lack of imaginaries, under a climate changed future.

Third, visuals of climate change are no different to visuals from those depicting any other contemporary issues, inasmuch as they are often problematic in terms of equity and justice. Media and journalism scholar Renita Coleman described how everyday news visuals can be 'laced with racial and gender stereotypes' (Coleman 2010: 235), in a way that would never 'pass muster' in the newsroom if they were written in text form. This is very often true for a full range of climate visuals – including and beyond news visuals – too. For example, as Chapter 2 shows, the proliferation of images featuring a lone Black child walking through a flood in a small island state bring along with it a set of ideas about how the world should work; ignoring the agency and power of indigenous island dwellers who are doing so much to shape international climate policy and practice, and innovating to safeguard their island futures. Climate visuals can marginalise and exclude some of the most vulnerable people, places and non-humans; perpetuating intersecting and ongoing injustices around power and (lack of) access to power, including around gender, race, class, age, poverty and legacies of colonialism.

About this book

This book brings together a diversity of interdisciplinary literature, my own research over the last 20 or so years as a scholar in visual climate communication, reflections from key figures in climate visual communications, as well as new research findings. As an environmental social scientist, I use a mix of methods; and this book reflects that. Both my existing research, and the new research presented here, variously uses methods including content analysis, critical visual discourse analysis, interviews, Q-method and focus groups. More recent work has increasingly been carried out collaboratively with computational social scientists, integrating approaches including computer vision, natural language processing and network science. The conceptual approach which underpins how these different methods are utilised and synthesised though is through framing theory (see earlier discussion on framing).

The body of this book contains five chapters which focus on five substantial areas of climate change action and enquiry. Each chapter then focusses on two different case studies, with each visual case study chosen for its prominence in the climate debate. Chapter 2, on climate adaptation, explores how we might adapt to manage the risks posed by climate change through the case studies of sea level rise and heatwaves imagery. Chapter 3 explores how climate impacts are realised on the world around us explored through a case study of that most iconic of climate visuals, the polar bear; and also through a study of flood images, also an everyday kind of climate impacts visual. In Chapter 4, on climate and energy, there is an examination of wind turbines and smokestacks images, and how these stereotyped visuals may help or hinder the fundamental challenge we face in the energy transition, in moving away from using fossil fuels. Chapter 5 focusses on climate science visuals, as science has been one of the most forefront ways of knowing (and disagreeing) about climate change (Hulme 2009). It tells the story of the emergence of that most viral of climate science visuals, the Climate Stripes; and it also explores how controversy has emerged over the changing of colour schemes on weather maps to be able to graphically represent our warming world. The final empirical chapter, Chapter 6, examines how people are visualised (or excluded) from climate imagery, through an analysis of climate images featuring politicians and climate protesters. The book's structure is deliberate, with each chapter able to stand alone, but also readable as part of a larger story about climate visuals. More specifically though, it is a deliberate decision to upend structures entrenched through authorities such as the IPCC (Intergovernmental Panel on Climate Change) and place, for example, the adaptation visuals chapter before the science visuals chapter; as a deliberate decision to challenge the epistemic authority assumed by science as the primary way of knowing climate change (see O'Neill and Hulme 2009).

Throughout, but particularly in the synthesis of the concluding chapter, this book seeks to challenge and critically question visual portrayals of climate change; and to suggest ways to work towards a more inclusive and responsible, but never neutral, climate visual discourse. The concluding chapter discusses the flow and friction of climate images as they move through a globalised media ecosystem. It first recognises that actors – from organisations, to groups, to individuals (including news organisations, image agencies, and even the images we hold in our own minds) hold substantial power to shape visual representations of climate change. And second, it explores how information architectures (that is, the way in which information is organised and presented, such as through search functionality and platform design) also hold considerable power to shape the climate visual discourse.

Thinking about the work of images in the world builds on a wide body of scholarly work; from art critic John Berger, cultural theorist Stuart Hall and novelist Susan Sontag among many others (see Berger 1972, Hall 1973, Sontag 1977; and the collection of essays in Evans and Hall 1999). Art and photography scholar David Levi Strauss' words are important, in thinking through the power and performance of images:

> it is images that are making the political changes that we see happening around us possible. Focusing on what those images are, and how they work, and how they can be changed to work otherwise, is not a side issue anymore, it's a necessity ... It matters what we imagine to be possible. Change can only happen if we imagine things differently. (Strauss 2020)

Inspiration also comes via the more recent discussion within Visual Politics scholar Roland Bleiker's (2018) edited volume on Visual Global Politics, and particularly by Political Anthropologist Nayanika Mookherjee's work in a very different context (that of memorialising wartime sexual violence). Using her words, this book intends to 'explore the social life of images' and see how, when intertwined with other perspectives, they come 'to perform or co-construct a global politics' (Mookherjee 2018: 208). Finally, this book has been written as AI really comes to the fore in media, journalism and beyond. Here, scholarly work around platforms, travel and friction of images in a global, digital media ecosystem (Rose 2016) has been important; as has research on emerging issues such as generative-AI for imagery and visual mis- and disinformation (Thomas and Thomson 2023, Thomson et al 2023).

It is important to raise here that the intention of this book is not to condemn any particular images and their makers or readers. Images move through a complex media ecosystem; where there is rarely a straightforward or linear progression between images, their creation, circulation and (re)interpretation. However, there is an aim to encourage readers to question how we visualise climate change. How does climate change come to be visually represented? Why do certain types of visuals come to dominate, and others get marginalised? What work might a wider visual discourse of climate change do in the world? And, importantly – if the climate visual discourse is limited (as this book argues), who has the power to act on these visual representations, in which ways, and what are potential points of intervention for opening up the climate visual discourse?

There are, of course, limitations to the discussions presented in this book. The most obvious is the focus on case studies arising mostly from the Global North, and the UK in particular. This is representative of the scholarly literature on climate communication, which has been more narrowly focussed on Anglophone and Global North nations (Moser 2016), especially for visual climate communication. Scholars, journalists and others seek to challenge this, exploring contexts in the Global South (for example, Takahashi et al 2021, Ejaz and Najam 2023, Veneti and Rovisco 2023, Carbon Brief 2024). Another limitation is due to my background as a scholar whose work has particularly focussed on the role of the news media thus far; although an increasing range of collaborations seeks to explore images in a range of other settings besides news, such as in TV soap opera dramas and on cookery shows, and in the advertorials shown on Facebook. This is important, considering that news avoidance is particularly common among some groups (for example, in the US, 12 per cent on the political left but 64 per cent on the political right say they actively avoid news about climate change; Newman et al 2023). Finally, although this book features a diverse range of visuals – cartoons, film posters, artistic illustrations, scientific graphs, weather maps, memes, campaign images and adverts – photographic images do dominate.

This book tells stories about climate visuals. But it is also inevitably a selection of stories about people and the work they perform: scientists, artists, photographers, image agency creatives, graphic designers, weather forecasters, journalists, activists … many people are involved in visual meaning-making about climate change. The book purposefully draws on some of their words to help put the visuals in context. This book also features many of the visuals which are discussed in the text, in colour – not as easy a task as one might imagine, in the world of academic book publishing. I hope you will engage in the storytelling performed by these images as much as you might with the words written about them.

2

Adaptation: Heatwaves and Sea Level Rise

From the earliest days, people have made decisions about the weather that they experience: is it going to be a hot or cold, wet or dry day? And, people have tried to figure out what the weather has in store for them for the following days, seasons and years: is it looking like a fairly settled period of expected weather, or might you be anticipating a heatwave, cold spell, drought or flood? Adapting to climate change is concerned with thinking about likely weather over longer time frames, and the significance of what those changes might be (Adger, Lorenzoni and O'Brien 2009). The IPCC (2022) defines adaptation as: 'the process of adjustment to actual or expected climate and its effects, in order to moderate harm or exploit beneficial opportunities'. This chapter is concerned with imagery about climate adaptation. Adaptation imagery is important because it reflects and shapes how we think and feel about living with climate risk, and how we might live with climate change into the future. This chapter explores two case studies: images associated with heatwaves in western European nations, and imagery associated with climate change in South Pacific small island nations. These case studies show how, when particular images about climate adaptation come to dominate, they can act to marginalise particular voices and to foreclose particular futures. It also offers some ways forward, by suggesting how images can offer a more equitable, just and representative narrative for climate adaptation.

Heatwaves

Adapting to extreme heat is an important part of climate change adaptation. Extreme heat events, or heatwaves, are projected to increase in occurrence, length and intensity due to climate change (IPCC 2022, Romanello et al 2022). Adapting to heatwaves is a significant public policy concern, with extreme heat carrying substantial health risks to both human mortality and morbidity. Health risks from heat exposure include heat stress, heat stroke and dehydration; but also (perhaps less visible) risks including impacts on mental health, an increased risk of accidents, and an additional health burden for people with chronic illness (WHO 2018). These risks are experienced unevenly. Those most at risk of extreme heat include older people, children, people with pre-existing medical conditions, those who are pregnant, work outdoors or are manual labourers, and those who live in poor-quality housing. The WHO states that these health impacts are largely preventable with specific public health actions (WHO 2018). However, communication of the public health risks of extreme heat can be challenging, as the health risks remain largely invisible (Brimicombe et al 2021) and those most at risk may not feel personally vulnerable (Abrahamson et al 2009). The communication of extreme heat is additionally challenging in places with cultural climate histories which tend to welcome a forecasted period of hot, sunny weather (Fox 2005).

I write this living in the UK; with its cool winters, warmer summers and often wet weather (Met Office 2023). Fox (2005) discusses the 'weather talk' of the UK, and how the changeability of the weather means a cultural preoccupation with talking about it. Even in the UK though, extreme heat has led to increased mortality. In 2022 the UK's Health Security Agency and Office for National Statistics estimated that there were 2,985 excess deaths due to extreme heat (UKHSA/ONS 2022). However, the visual representation of heat extremes often fails to convey this risk. During the hot summer of 2019, I wrote a guest post for the blog *Carbon Brief*, titled 'How heatwave images in the media can better represent climate risks' (O'Neill 2019). It was the impetus to carry out empirical analysis of media visual representations of extreme heat. With a team from four European countries – the Netherlands, Germany, France and the UK – we collected and analysed the images published by 20 online news organisations, in stories featuring both the terms 'climate change' and 'heatwave' (with translations) during summer 2019. This resulted in an image dataset of 245 images. We examined the media reporting in these news articles: both in the text, and also the lead visual attached to each of these stories. Note that, in many cases, restrictions around copyright prevent the use of mainstream news images being reprinted in this book; and so Creative Commons licensed alternatives have been used to illustrate findings. The dominance of the trends discussed next can also be evidenced through viewing the montage of UK newspaper frontpages collected by journalist organisation Carbon Brief after the UK reached a record-breaking 40°C in summer 2022 (see Carbon Brief 2022; again, their montage cannot be reprinted here due to copyright issues).

'Fun in the sun' and 'the idea of heat'

Our findings showed that many (31 per cent) heatwave visuals were positively valenced. This meant, in our coding scheme, that the image presented a

Figure 2.1: Typical 'fun in the sun' heatwave image

Source: Funk Dooby at Flickr, CC BY-SA 2.0

heatwave as being a positive event (something fun, enjoyable, holiday-like, or relaxing). In contrast, article texts were very rarely positively valenced (<1 per cent). So, while the text of these heatwave articles was often consistent with public health messaging around coping and adapting to extreme heat events, the images were not.

In all four countries, the most prevalent type of images were those showing people having a fun time in or by water. We called this dominant framing 'fun in the sun'. It is exemplified by images of people splashing in a city fountain, or enjoying sunny weather at the beach. Figure 2.1 is an image typical of this 'fun in the sun' visual frame. This image shows many people swimming and sunbathing at the beach. Together with the brightly coloured and saturated nature of the image, it connotes the feeling of a heatwave as a summer holiday-like event, rather than a heatwave as something that is potentially risky. This image has accompanied at least two heatwave articles, which are typical in their use of positively valenced beach imagery against negatively-valenced text: 'As Europe Suffers "Heat Apocalypse," UK Smashes Temperature Record' in the Earth Island Journal (Earth Island Institute 2022) and 'UK, Germany, France on Pace for Their Hottest Year on Record' from the Yale Environment 360 website (2022). It is notable that both these articles are published by organisations promoting environmental journalism and advocacy, demonstrating how entrenched and accepted the 'fun in the sun' trope is, even in organisations writing about the dangerous and unequal impacts of heat extremes.

There was another common visual framing, which we called the 'idea of heat'. While this did connote heat and danger, for example, using bright sunbursts, saturated colours or 'dangerous' and 'hot' red or orange colours (see Figure 2.2 for an example image; see also Chapter 4 for an in-depth discussion of colour and hue), people were largely absent. Where people did

Figure 2.2: Typical 'idea of heat' image

Source: Ugurhan, iStock

feature in the 'idea of heat' images, they were often pictured alone and were often depersonalised by obscuring identifiable facial features – for example, people's faces were pictured drinking from a water bottle silhouetted against the sun, so their faces were cast into deep shadow. Figure 2.2 shows a typical example of the 'idea of heat' visual frame, featuring an analogue thermometer giving a reading of at least 45°C, a generic urban skyline, and an orange sky and sunburst. The thermometer reading, orange hue and sunburst all connote heat extremes. The buildings in the background are the only indicators of human impacts from this heat, as people are entirely absent.

Our final finding concerned the difference between the meaning constructed by the text of a news article, and the meaning constructed by the article's visuals. We found that the dissonance between article texts and images could be very stark. For instance, examples in the dataset show headlines describing unprecedented dangerous heat events, the experiences of vulnerable people, and even fatalities due to the extreme hot weather, alongside holiday-like photos using the 'fun in the sun' visual framing. This echoes a Canadian newspaper study carried out by DiFrancesco and Young (2011), where they found climate news article content could pull in opposite narrative directions, with the text of a news article and its associated image making unrelated and sometimes even contradictory claims; perhaps due to different responsibilities regarding text and images within the newsroom environment falling to different teams; and subsequently different perceptions of a news story and news values.

Heatwave visuals marginalise vulnerable people

There are two problems raised by this current visual framing of heatwaves in the northern European news media. The first is that, by giving so much space to the 'fun in the sun' visual frame, the experiences of those vulnerable to extreme heat events are almost completely excluded. By marginalising the experience of older people, young children, outdoor workers, and those in low quality housing, for example, the very real concerns for people's health and wellbeing during extreme heat are lost from the public discourse. Indeed, global polling of news media audiences indicates the UK is particularly notable for the low percentage of the population who consider that climate change is having an impact on either their own, their family's, or their country's health (Ejaz, Mukherjee and Fletcher 2023). The dominance of 'fun in the sun' is also a missed opportunity for raising the profile of public health messages about who is vulnerable, and how vulnerability to extreme heat can be reduced. Public health researchers have shown both that chronically ill elderly people (that is, those at higher risk of extreme heat) do not perceive themselves as personally at risk from heat extremes (Wolf, Adger and Lorenzoni 2010); but also that media storylines could usefully portray these risks, as an additional and powerful communications channel alongside traditional weather warnings (Abrahamson et al 2008).

The second problem raised by this visual discourse of extreme heat is that the dominance of 'fun in the sun' excludes opportunities for imagining a more resilient future. Even where images do provide visual cues for behavioural adaptations to extreme heat – such as in the featureless, shadowed faces of a person drinking from a water bottle in 'the idea of heat' visual frame – these images' extreme blandness and likely effect of distancing the viewer from the issue, are far from good practice on how to engage people with climate risk (see Corner, Webster and Teriete 2015).

Beyond 'fun in the sun' for heatwave visuals

A more engaging visual discourse is possible, though. It is certainly not self-evident that 'fun in the sun' is the only way to visualise extreme heat events. In our study, there was one media organisation that was an exception in the way it pictured heatwaves: the Dutch daily newspaper, *Algemeen Dagblad* (AD). AD used images consistent with climate solutions journalism. For example, they used a photograph by Theo Peeters showing support workers handing out ice creams to elder care centre residents (Barendregt 2019), effectively challenging the 'fun in the sun' visual frame. An additional article pictured a two-way slider image of a grey, urban communal city space transformed with planting to reduce the urban heat island effect; offering a visual depiction of how to adapt urban areas to the increasing risk of extreme heat.

Figure 2.3: Alternative heatwave imagery, depicting a visibly uncomfortably hot Black delivery driver taking off his glasses and facemask to wipe his brow

Source: Tom Williams/Associated Press/Alamy Stock Photo

Other media organisations have also used alternative visuals. The BBC news website featured a Black delivery driver wiping his brow and looking visibly uncomfortably hot (Figure 2.3) to accompany the article 'UK heatwave: Temperatures to hit low 30s as heat-health alert issued' (BBC 2022). This image is an example of how to visually portray the uneven impacts of extreme heat events: heatwaves disproportionately impact manual labourers, for example (WHO 2018); and lower skilled job roles are more likely to be held by Black workers (ONS 2022). This image also demonstrates the international flow of news images. It is a stock image from Getty Images depicting Amazon driver Shawndu Stackhouse. Sharp-eyed readers may have noticed how he is sitting in a driving seat on the left-hand side, rather than the UK's right; as this is an image of Stackhouse delivering packages in Washington DC rather than a UK-based driver.

Work from beyond the northern European context also provides examples of how to open up the visual discourse of extreme heat. I have been working with colleagues Constantine Boussalis, Ranu Malla and Travis Coan, all computational social scientists, to analyse heatwave visuals from around the world. This emerging work indicates that a different visual discourse of extreme heat is in evidence elsewhere in the world. We carried out a study of news stories and their images using both the words 'climate change' and 'heatwave' (and their translations) from diverse media sources across 14 different countries; analysed using computer vision approaches.

The emerging results are notable for their difference between more climate vulnerable countries (broadly, countries in the Global South) and countries relatively less vulnerable to climate change (broadly, countries in the Global North; see ND-GAIN 2024). Global South countries have a strong visual discourse portraying the impacts of extreme heat on vulnerable people, and the 'fun in the sun' visual frame is less common (Boussalis et al 2024). This could be because countries like India have a tropical monsoon climate, where extreme heat has been a significant threat historically, and which is intensified under climate change. However, countries like Australia and the USA, which also have historical experience of extreme heat events, do still use the 'fun in the sun' visual frame. Further research could usefully investigate these examples, and seek to understand how the cultural construction of a heatwave influences its media portrayal and any subsequent public health impacts.

There may be structural, routine-based newsroom explanations for why bright, saturated images are consistent features in the reporting on heatwaves. Alastair Johnstone, Deputy News Picture Editor at the UK's Times newspaper (2018–2022) explains (personal communication, 5 January 2024):

> Often, an editor would seek out bright, cheerful, colourful images to accompany heatwave reporting. These were the sorts of images they imagined viewers wanted to see in a newspaper in the summer time. Also, aesthetically, we were thinking about balancing the newspaper layout: the darker sections [of black text] against the [brighter, more colourful] images. Flood images, for example, were often considered 'too brown'.

Other newsroom structures may also contribute to the dominance of 'fun in the sun' images for heatwave news. An impending heatwave might start as a story on the science or environment desk, but transfer to the general news desk if the heatwave is predicted to be particularly extreme, that is, if it starts to becomes a more newsworthy story. While perhaps specialist climate reporters may be cognisant of visually reporting the dangers of extreme heat events alongside text reports, this nuance may be lost as the story transfers to general news (see Strauss et al 2021).

It is important to be clear that this is not a call to redact all images of people enjoying sunny weather on a hot day. There is not necessarily a clear distinction between enjoyable, warm weather and extreme heat – sometimes this distinction is very obvious, but other times less so. For example, different countries have different definitions of what constitutes a 'heatwave', dependant on their context (see Met Office 2023 for the UK's definition). These indicate that the decision of when it might be appropriate to use 'fun in the sun' images is not necessarily straightforward. As BBC Weather Presenter, Ben Rich, comments (personal communication, 30 June 2023):

This is the sort of subjective judgement call we have to make all the time. It's all about context -- a week of warmth and sun after months of rain is probably 'good' news for most people. 40C and drought ... not so much. We try to adjust our storytelling and visuals to match, but it's a tough balance!

What does seem to be clear is that the dominance of 'fun in the sun' images to the exclusion of others, especially when juxtaposed to text talking of the substantial risks of extreme heat events, is problematic. A critical eye on heatwave imagery is an important part of dialogue on how to adapt to increasingly dangerous heat events and their unequal impacts.

Sea level rise

The IPCC has a chapter devoted to understanding impacts, adaptation and vulnerability in small islands (Chapter 15 in IPCC 2022); where the geographical remoteness and isolation, small land area, low elevation and freshwater sources, among other similarities, are discussed in terms of the substantial exposure and sensitivity small islands face in regard to climate change. There is far less acknowledgement, or integration, of the role of the dynamic land-ocean relationship, which challenges how the accepted scientific wisdom of global climate modelling plays out at the level of atoll dynamics and vulnerability (Kench, Ford and Owen 2018). These western perspectives also ignore indigenous Pacific islanders' understandings of their territory where there is so much more to the region than land surface – there is a 'sea of islands' and not 'islands in a far sea' (Hau'ofa 1994). The idea of island 'smallness' is something constructed by western colonisers, and not something recognised by island dwellers. As Tuvaluan Taukiei Kitara explains:

> We see ourselves as not small island states, but large ocean states ... we include land and sea together. They're both in one. To tell that connection between land, people and the sea, I'll tell a story about myself: when I was born, my umbilical cord was cut in two. One [part] is buried on land and the other half of my umbilical cord was taken out by my dad to the ocean and was thrown into the sea. That tree, my grandfather said, you know, you recognize that tree, it's a coconut tree, that is you. Underneath that tree is your umbilical cord. So you are the custodian, you have responsibility to the land. But the other half of your umbilical cord is in the ocean - you also have a duty and responsibility to the ocean. And so we do not distinguish between land and our seas, because these two things coexist. We do not differentiate them. (Taukiei Kitara, independent scholar and indigenous knowledge holder, Tuvalu, personal communication, 10 June 2024)

These insights are of essential importance for understanding the vulnerability of small islands, as timescales and probabilities play a key role in understanding potential adaptation options: for example, the high risk of maladaptation for small island dwellers facing resettlement in anticipation of potential climate impacts compared to lower risk, higher reward adaptations such as voluntary labour mobility (Barnett and O'Neill 2012). Moving beyond scientific knowledge, there is still less recognition or integration, at the global scale, of how people living in small islands are already well adapted to these challenges, through their indigenous knowledge and governance systems (Farbotko and Campbell 2022). If, as Jarillo and Barnett (2022: 849) state, the 'issue of moral concern is the habitability of atolls', then such knowledge is of fundamental importance for understanding habitability risks (Farbotko and Campbell 2022). Yet, as the following images tropes of climate change and small islands show, currently ubiquitous visuals tell a story symbolising inevitability and smallness – the sliver of land, the lone child – both powerless in the face of a malevolent ocean.

Aerial imagery of coral atolls

Figure 2.4 is an example of a common visual trope illustrating climate change in atolls. The image is a photograph incorporating both tropical ocean and land; here, showing the capital Funafuti in the atoll nation of Tuvalu in the South Pacific. This is perhaps not surprising: if sea level rise is the climate impact of concern, what photo might one expect, but an image comprising both sea and land? Images like Figure 2.4 also bring aesthetic qualities important for a newsworthy image: here, the curving, linear island shape, deep colour saturation, and contrasting rich colour hues fill the frame. Unpicking this image, though, brings a different perspective. These ocean-land images are often aerial images. Aerial images explicitly visually narrate the juxtaposition between the vast ocean against the small land area of all coral atoll nations. The perceived fragility – from a 'Western' gaze – of this narrow strip of land stands in direct contrast to the forbidding and dark ocean, lapping at its shores. This is not the perspective of island inhabitants, where the land-ocean relationship is instead summed up by the words 'we are the sea, we are the ocean' (Hau'ofa 1994: 160); the ocean is neither threat nor foe in contrast to land, but both together form an integral part of identity and everyday experience.

In such images, buildings are often rendered imperceptible and people are certainly not visible. In this particular image, houses are just evident in the centre bottom of the image; although one's eye is drawn away from them by the linear island snaking towards the vast, distant horizon at the very top of the photograph. No people are visible. This is a 'God's eye' view, surveying the land from up high and from a distance. There is a problematic

Figure 2.4: Typical aerial view imagery to depict sea level rise threat to coral atolls

Note: Original caption: 'Rising Sea Levels Threaten Coral Atoll Nation of Tuvalu'
Source: Mario Tama, Getty Images (2019)

history of satellite imagery and aerial photography positioning the viewer as an 'overseer', with authors such as Mirzoeff (2011; see also Spiegel 2020) critiquing these images in terms of colonial histories of slave plantations, with implications for how territory is represented and perceived. It is possible that this is exactly what some of these images are: photographs taken while on the plane itself, as the travelling photo-journalist flies quickly in and out 'on location', to countries distant from their own geographical homelands; although the quality of this particular image and this photographer's past practice suggests the use of a drone (Schneider 2021).

This image was taken by Mario Tama, an award-nominated American photographer. He is based in Los Angeles and travels extensively as a staff photographer for the global image agency Getty Images (Taylor 2017). Just as this image was created and shared within a globalised media ecosystem, so also the image search engine TinEye finds this photograph in use many times across online news media; including in an Irish news story reporting on a poll for climate action (Digital Desk Staff 2021), a *Washington Post* article on international climate diplomacy (Birnbaum and Kaplan 2022), and Tuvalu's plans to build a digital replica nation in the Metaverse, in the Sydney Morning Herald (Craymer 2022). This image has travelled as a cosmopolitan imagination of what climate change is in small islands. What is depicted is a western imagination of futures in coral atolls and, by extension, to all small islands. What is erased from this image, and island atoll aerial images like this, is the role that island dwellers play in living in and living with the land-ocean. It encapsulates the 'wishful sinking' climate discourse (Farbotko 2010: 47) of the western media, where islands are only of utility when they disappear beneath the waves, as a visual harbinger of the truth and urgency of global climate change (for a recent example of how ingrained yet futile the 'wishful sinking' narrative continues to be, see the example of novelist Kathie Lette joking about Tuvalu drowning, while miming putting a snorkel on and giving a royal wave during primetime TV coverage of the UK's coronation celebrations; Boyle 2023). This image, and images like this, continue to project an eco-colonial gaze on Tuvalu and other small island nations.

Young girl in floodwater

Figure 2.5 is an example of another visual trope common to the visualisation of sea level rise in small islands (for example, Sakellari 2021). Here, a young Fijian girl, who the caption names as Miriam Bulivono, walks through a shallow, flooded grassy area. Although there are people in the far background, she is on her own, without accompanying friends or family. One-story houses are visible in the background, and everyday domestic signs such as a clothes line full of washing, also show the flood lapping at

Figure 2.5: Typical sea level rise imagery, featuring a young Tuvaluan girl

Note: Original caption: 'An indigenous Fijian girl (Miriam Bulivono age 8) walks in her village on flooded land in Fiji. On Feb 2016 Severe Tropical Cyclone Winston was the strongest tropical cyclone in Fiji in recorded history'.

Source: Rafael Ben-Ari, iStock (2016)

their edges. The young girl walks on, with her head bowed. This image has much in common with historic (and current) humanitarian aid imagery and its critiques. These pictures feature women and children, often looking pensive or subdued, photographed outside and alone, away from the safety of physical structures such as homes (Briggs 2003, Manzo 2008). Images such as these construct a portrayal of innocence, neutrality and solidarity; but also act as a visual construction of the need for intervention by a powerful overseer.

As with the aerial image of a coral atoll, this photograph of Miriam Bulivono also demonstrates how island imagery travels through a globalised, highly networked media ecosystem to represent a much wider set of issues, that is, it becomes a climate metonym. (Here, a metonym refers to the relationship between a representation and the way in which it works in the world; in particular, around how part of something comes to represent a much wider set of ideas and relationships; see O'Neill 2022). This photograph was taken on 18 December 2016, by Rafael Ben-Ari, an Israeli-born, Australian-based travel and war photographer. It was uploaded on 25 January 2017 to stock image libraries including iStock, Adobe Stock, Alamy and Shutterstock. On iStock, this image sits in Ben-Ari's web collection of Fijian travel photography (iStock 2023), looking somewhat out of place besides all the other photographs which use stereotyped visuals of a tropical island paradise – snorkelling over coral, exclusive island resorts, delicious food and traditional dancing – common to glossy magazines and Instagram aesthetics (see Leaver, Highfield and Abidin 2020). These stock libraries offer cheap access to images: at the time of writing, this image is available for licensing at the cost of £7 (or cheaper with a subscription) on iStock. The image was uploaded with an 'editorial only' licence, meaning the image does not have any model or property releases (permissions), so it can only be used in connection with content that is newsworthy or of general interest, rather than commercial or advertorial based uses.

The caption describes the image as: 'an indigenous Fijian girl (Miriam Bulivono age 8) walks in her village on flooded land in Fiji. On Feb 2016 Severe Tropical Cyclone Winston was the strongest tropical cyclone in Fiji in recorded history' (iStock 2023). So, the cyclone hit in February 2016, the photograph was taken ten months later, and the connection between Cyclone Winston and Miriam's image was only made when the image was uploaded to stock image libraries in 2017, almost a year after the cyclone happened. While the image depicts a young girl alone in standing water, and the caption referenced a tropical cyclone, this is not an image of someone experiencing Tropical Cyclone Winston. Yet captions state the photograph is of Cyclone Winston, for example 'An Indigenous Fijian girl walks over flooded land after Cyclone Winston in 2016' (Kaminski 2023; see also Slezak 2017).

Figure 2.6: The 'wishful sinking' media narrative, Tuvalu. Tuvaluan children have fun swimming in the sea, as part of a child's birthday party gathering in Funafuti. Two filmmakers stand at the edge, filming them for a British documentary.

Source: Reprinted with permission of Carol Farbotko (2023)

In Tuvalu, Farbotko has evidenced a two-decade long history of a veritable flood of media actors – foreign journalists, researchers, environmentalists and documentary-makers flying in to recreate expected climate visual narratives. Photojournalists and others come to Tuvalu to film king tide flooding, recreating a visual discourse of local children wading through floodwaters, while the image-makers stand their cameras on land just out of the frame and fly out as the tide recedes (see Figures 1 and 2, taken in 2006, in Farbotko 2010); capturing a generalised visual narrative of island life to match the expectations of the 'wishful sinking' global discourse (Figure 2.6, taken in 2023).

The earliest use of the Miriam Bulivono image found on a news site was in May 2017. Similarly to the aerial shot in Figure 2.4, reverse image search engine TinEye suggests the Miriam Bulivono image has been reproduced in news sites many times since. It is notable, in this case, how this image has been used so much by one particular news site, UK newspaper *The Guardian*. The image has been used to illustrate at least seven climate news stories in *The Guardian* between 2017–2023: from reporting on Trump and the Paris agreement (Milman, Watts and Phillips 2017), to Fijian adaptation funding (Slezak 2017), climate change as a security threat in Asia Pacific (Doherty 2020), Australian-Pacific migration policy (Doherty 2020), to the rights of children in the face of climate change (Kaminski 2023); as well as a comment piece on indigenous Pacific activism (Fruean 2021). In many cases, this image stands in for a much broader visual discourse of 'small islands'. In one case, the caption and article (Sparrow 2019) do not even mention Fiji or the Pacific; using the photograph to presumably stand in for a set of unspoken assumptions about climate vulnerability.

Perhaps this should not be surprising: newsroom routines related to cost-management may lead to certain images being used (and then being used again and again). This could include the subscription a news organisation has to a particular image agency (an editor is then likely to preferentially use this image agency to select an image from, as their pictures have already been effectively paid for through an existing subscription, rather than needing a separate image fee to be paid for image rights). Although unlikely the case for this image of Miriam Bulivono (as it is under an 'editorial' licence), the licensing arrangements of stock imagery (like the staged thermometers and sunbursts of Figure 2.2) can allow unlimited use of the same image, once the one-off image rights have been bought. Additionally, newsroom routines related to how images are archived, once they have been used by a publication, may increase their chances of reuse. For example, UK newspaper *The Times* uses a form of content management system that makes images which have already been used and archived more visible to users, and thus perhaps more likely to be subject to reuse (Alastair Johnstone, Deputy News Picture Editor, *The Times*, 2018–2022, personal communication, 5 January 2024; Hayes 2024). There may be other aesthetic or editorial reasons that *The Guardian* news staff have reused the Miriam Bulivono image. However the image has come to be used so much, what is clear is that this sort of island imagery has come to act as a synecdoche for climate change: the photograph simultaneously easily connotes something (viewers can easily 'read' this climate image), but also, such images come to mean very little at all (this image trope comes to 'slide towards radical interchangeability'; McQuire 1997: 59). A child walking through a shallow flood in Fiji comes to represent overarching global climate risk to powerless 'others'.

Sea level rise imagery removes agency of island dwellers

These image tropes are deeply and seemingly intractably entrenched. During COP26, *The Guardian* printed a powerful opinion piece by the Samoan climate activist Brianna Fruean and representative of the Pacific Climate Warriors delegation, titled 'Pacific islanders aren't just victims – we know how to fight the climate crisis'. *The Guardian* is notable as a global media organisation which has made a public editorial commitment to rethinking the images used for its climate journalism (Shields 2019); and here was a piece looking to give voice to indigenous Pacific perspectives. Yet the image that accompanied Brianna Fruean's narrative on resisting climate colonialism? It was yet another reuse of the Miriam Bulivono image in Figure 2.5.

Similarly, even when Tuvaluan Foreign Minister Simon Kofe was nominated for a Nobel Peace Prize, media coverage focussed on his contribution via a speech to COP26, filmed while he stood at a lectern positioned in the sea; but failed to report fully on Kofe's other substantial and innovative climate leadership achievements which had also led to his nomination. The dominance of the 'sinking islands' visual trope acted to reimpose the global media narrative of Tuvalu as a climate change victim; even when the news hook was of Kofe's and Tuvalu's resilience, resistance and innovative adaptation for climate change (Farbotko and Kitara 2022).

Pacific islanders' challenge dominant visual narratives

What are the alternatives? As Jarillo and Barnett (2022) argue, challenging the dominance of the western focus of the threat to islands from sea level rise is not the same as downplaying the very substantial risk of sea level rise to island dwellers. And, challenging these visual portrayals is intended to be constructive and not combative: many in the global media ecosystem are likely well-intentioned in their efforts to visually portray the urgency of climate change as a presumed starting point for climate action (even if this presumed, straightforward link from fearful visual to climate action is predicated more on hope rather than evidence; see O'Neill and Nicholson-Cole 2009). Indeed, an interview with photo-journalist Mario Tama (photographer of the aerial shot in Figure 2.4, though the interview was not in relation to that particular image) states that he believes climate change is 'the story of our time' and that photographers like him are 'obligated … to bring to life all of the pieces of this giant puzzle of our warming planet' (Tama, quoted in Schneider 2021).

However, visual tropes such as the aerial view of the coral atoll and the lone child in floodwaters act to construct Pacific nations as small, inert and passive in the face of inescapable climate change. They focus exclusively on the threat of sea level rise while exerting an eco-colonial gaze (Farbotko 2010). Non-indigenous perspectives set the visual tone of a 'dangerous' ocean encroaching on the 'safe' land; and construct the ocean only as a 'menace' (Jarillo and Barnett 2022: 851). Island dwellers are scripted into a western narrative of victims fleeing inundation, perhaps even already drowned; removing agency. As Carol Farbotko explains (see also Farbotko 2010):

I know there's a lot of expectations on journalists to tell the story to the audience that they serve. So, you know, if it's a Western news service, it's pitched at Western audiences and journalists are often told that they need to frame the story in a way that will make sense to that audience. And, often that is where the problem starts. So there's issues like the victim narrative that's perpetuated no matter what, because that's going to serve this audience, because that's what they expect to see. The developing world is a victim of climate change and there's no other possibility. (Carol Farbotko, Geographer, Griffith University, personal communication, 10 June 2024)

This is very different to the perspectives of Pacific islanders', where the ocean is a connector of places and as much a part of a connection to home and place as land (Hau'ofa 1994, Diaz and Kauanui 2001); and where climate advocacy draws on an oceanic identity encapsulating ancestral responsibility to ocean, islands and home (Titifanue 2017). This is encapsulated by the Climate Warriors campaign imagery, which typically depicts Pacific climate activists in ceremonial dress of their diverse cultures, in battle-ready stance. Farbotko and Kitara (2022) describe this 'warrior aesthetic' as making use of ecological materials common across the Pacific, including coconut fibre, shells and pandanus; as well as the key concept of vaka (or canoe) to centralise values at-risk, but also hope of survival in a self-determined journey navigating climate change. This visual aesthetic is common where Pacific Island dwellers have control over their visual representation, such as in Pacific Climate Warrior social media accounts (350 Pacific 2023, Titifanue et al 2017, Farbotko and Kitara 2022). Here, the visual representations are far from Pacific islanders as climate change victims common to the western media; instead, these warrior aesthetic visuals seek to engender a unified Pacific identity, highlighting their solidarity and strength (Titifanue 2017).

Alternative visuals to the Pacific Climate Warriors also exist elsewhere on social media platforms, advancing a quiet visual resistance to the dominance of the climate victims' visual narrative (Carol Farbotko, personal communication, 4 July 2023). The Facebook page and Flickr account of the Tuvalu Coastal Adaptation Project (TCAP) is replete with images of this ambitious United Nations Framework Convention on Climate Change (UNFCCC) Green Climate Fund financed project, designed to build coastal resilience and manage coastal inundation risk in three of Tuvalu's nine inhabited islands. On the TCAP Facebook Page (TCAP 2023, Flickr 2023), images feature Tuvaluan schoolchildren learning about the project; Tuvaluans working on the TCAP project; and aerial shots which challenge stereotypes in showing both the TCAP land reclamation and construction, alongside the azure ocean and palm trees (Figure 2.7). Similarly to the

Climate Warriors images, these TCAP photographs work to challenge the climate victim narrative and instead represent a hopeful and liveable future for island dwellers; while not diminishing the scale and urgency of the climate challenge. These images, or similar visual themes, appear to be entirely missing from global media reporting of climate change in small island nations.

Where to next, for climate adaptation images?

This chapter has examined images about climate change adaptation. It has shown how dominant imagery about climate adaptation has the power to marginalise the voices and experiences of particular groups of people: from island dwellers in the South Pacific; to older people, young children, manual labourers and people already experiencing ill-health living in western Europe. Such images can act to foreclose options for adaptation (Farbotko et al 2023). In the South Pacific, the dominance of imagery which reinforces a western perspective of islands' smallness and powerlessness attempts to ignore the agency of nations like Tuvalu, who are working at the forefront of innovative climate politics and climate adaptation. In western Europe, the dominance of 'fun in the sun' heatwave imagery stands in contrast to public health advice intended to improve health and wellbeing outcomes during extreme heat events. But adaptation imagery also has the power to give voice and representation to

Figure 2.7: Alternative visions of Tuvalu adapting to the impacts of sea level rise

Note: This is an aerial photograph of the TCAP land reclamation project in Funafuti.

Sources: Tuvalu Coastal Adaptation Project (TCAP), UNDP Climate Flickr collection (Flickr 2023), James Lewis/TCAP, CC BY-NC 2.0 DEED

Note: Tuvaluan climate activist and scholar Taukiei Kitara rests in front of the TCAP land reclamation project.

Source: Reprinted with permission of Taukiei Kitara (2023)

often already-marginalised people: there are examples of both heatwave imagery and imagery of small islands that centre the lived experience, and essential local and traditional knowledges and values of those in place. Images can offer a more equitable, just and representative narrative for climate adaptation.

3

Impacts: Polar Bears and Flooding

Images showing the potential impacts of climate change are very common. Think of a TV news broadcast panning over a flooded cityscape, an award-winning photograph in an exhibition depicting a polar bear clinging to an ice floe, or a social media post showing a raging wildfire engulfing someone's home. Impacts imagery has long been used to try and engage people with climate change (Lester and Cottle 2009, Rebich-Hespanha and Rice 2016, O'Neill 2020). The IPCC (2022: 2912) defines climate impacts as: The consequences of realised risks on natural and human systems, where risks result from the interactions of climate-related hazards (including extreme weather/climate events), exposure, and vulnerability. This chapter examines two iconic visual tropes of climate impacts imagery: polar bear images and flood images. It shows how these visuals have been constructed in fairly consistent ways; and what the potential readings of these image types can be. The chapter concludes by critiquing this sort of impact-led, common and pervasive imagery, and suggests ways forward for visualising climate impacts.

The polar bear

Polar bears are a highly specialised species who depend on Arctic sea ice as a platform for travel and hunting (Derocher, Lunn and Stirling 2004). The rapid changes seen in the Arctic, and the Arctic's potential to be ice-free in summer before 2050 (Notz & SIMIP Community 2020) puts polar bears at significant risk (Derocher, Lunn and Stirling 2004). Polar bears were first listed as 'threatened' on the ICUN (International Union for Conservation of Nature) Red List in 2015, the first time a species was listed due to climate change (IUCN 2015). Polar bear imagery has become ubiquitous in climate communications. They appear on news bulletins, magazine covers, in films and documentaries, on protest placards. As early as 2007, they were called the 'poster child' of climate change (Garfield 2007); and polar bear imagery is now so associated with climate change that it means nothing but climate change (as well as meaning almost nothing about climate change either, as their ubiquity comes to blur earlier meanings). This section follows the evolution of polar bear imagery through the last 30 years, from Coca-Cola adverts to climate cartooning and memes, showing the growth of climate scepticism as bear imagery became subverted and parodied.

Polar bears are polar bears

In the early 1990s, polar bear population change was only linked to climate change by scientists. In the cultural sphere, polar bears were, at this point, still viewed as dangerous apex predators. This is exemplified by a series of photographs taken by photographer Norbert Rosing of a polar bear playing with a domestic husky. These images were published in the magazine *National Geographic* in 1994. However, readers did not view the polar bear in these images as cute and cuddly – far from it. Instead, Rosing was accused of putting the husky's life in danger by allowing it near such a dangerous predator (Krulwich 2014). Things began to change from the mid-1990s. Although Coca-Cola had used polar bears in some advertising as early as 1922, the brand really became closely associated with polar bear images from 1993 in the 'Northern Lights' advertisement series. In these adverts, polar bears were depicted hanging out in the Arctic, having a good time – and always enjoying a Coca-Cola. The association between cute, cuddly polar bear imagery and the brand has continued to the present day (Coca-Cola 2021).

In the UK and US print news media, ice imagery was common through the early 2000s; it reached as much as 25 per cent of the total visual climate coverage in the UK's *Daily Mail* and *Telegraph* during 2002. In contrast, polar bear imagery was only sporadically linked to climate change in UK and US print media during the early 2000s. From 2005, this situation began to reverse. Ice imagery declined in volume, and polar bear imagery began to increase. During this time, polar bear images were only used to accompany stories directly reporting on the decline of sea ice as a threat to the species (O'Neill 2019).

The sorts of images used by the US and UK news media during the mid-2000s look visually very similar to much polar bear imagery today: they show an adult bear, or a mother and cub, both pristine white against the pure white of the surrounding snow and ice (Figure 3.1). The presence of blue-grey water signals to the potential danger of melting ice. Depicting the Arctic as distant, vast and inaccessible, as well as a pristine wilderness, is well established in at least western cultures (recognising here the problematic colonial nature of terms such as 'wilderness' (Cronon 1996) as well as how these images largely erase indigenous inhabitants and their experience, for example, Huntington et al 2019). As Born (2018: 11) notes, humans are 'conspicuously absent'; with the white, inaccessible landscape reinforcing

Figure 3.1: Typical polar bear image

Source: Vladsilver, iStock

associations of transcendence (Cosgrove and della Dora 2008). Indeed, Cosgrove and della Dora suggest that these images present the Arctic as an 'eschatological' landscape, where 'apocalypse can be easily imagined' (2008: 5). These images, when accompanying a news story about climate change, also visually represent climate change as an issue that is distant in both space and time from the reader; that is, climate change is psychologically distant (O'Neill 2013; McDonald, Hui and Newell 2015).

It is always instructive, when reading images, to consider what alternative visuals might depict. A reasonable critique of this analysis is that polar bears are solitary species, who spend much of their time on ice. Of course images of polar bears would depict these features! Indeed. However, note how these images do not show, for example, the considerable speed, power or size of these species as apex predators. These polar bear images do not interrupt the vast white landscape with the red blood of a seal kill, for example; nor do they depict bears catching or killing prey; and they do not depict the millennia of Arctic indigenous people's habitation of their environment.

Polar bears mean politics

Public attitudes towards polar bears were shifting by the mid-2000s. Recall those photos taken by Rosing Norbert in 1991, and which caused much public anger when pictured in *National Geographic*, for potentially endangering the life of a husky by allowing it close to a dangerous polar bear. They came to light afresh in 2007, when they were put online as part of a radio show. Public reaction was completely different, as people loved the friendly, cuddly, huggable bear (Krulwich 2014). But as well as being cute and cuddly, bear images also started to be subverted and used in parody. Public polling was also starting to show increasing levels of climate scepticism (Capstick et al 2015). UK news images began to ridicule the idea of climate change impacting polar bears through, for example, an extreme close-up of a winking polar bear photo captioned 'sceptical: a polar bear weighing up the evidence for climate change' in *The Telegraph* in 2007; and a photograph of a huge polar bear perched atop a tiny ice floe to illustrate the *Daily Mail* stories 'Don't panic (much)' in 2008 and '10 mad ways to save the planet' in 2009. Polar bear imagery had become so 'tired and hackneyed' that the UK's *Guardian* newspaper avoided it wherever possible (Leo Hickman, features journalist and editor, *The Guardian* (1997–2013); personal communication, 13 July 2018).

There are several explanations for the politicisation of polar bear imagery from the mid-2000s onwards, in both cultural and formal politics. In 2005, there was a call to list polar bears under the US Endangered Species Act (ESA). In 2008, and after significant controversy, polar bears were listed as 'threatened' under the ESA (CBD 2018). While some applauded the ESA listing as a potential path for legal action for climate mitigation, President George W. Bush stated in a Rose Garden speech that this listing was 'never meant to regulate global climate change' (Bush 2008: np). Polar bear visuals crossed from formal to cultural politics through the life of orphaned polar bear cub, Knut, born in captivity in Berlin Zoo in 2006. In a phenomenon dubbed 'Knutmania', the polar cub's likeness appeared on the cover of Vanity Fair in 2007 and on a stamp in 2008, as a mascot for climate change via the German Environment Minister (Engelhard 2016). Alongside these events, a clip of a polar bear struggling, and failing, to climb on to a tiny ice floe in a vast ocean, was part of the film *An Inconvenient Truth*, in part a high-grossing Academy Award-winning documentary, but also a political campaign to bring about climate action via the Climate Leadership Corps campaigner training (Pearce and Nerlich 2018). As polar bear imagery became embedded in formal and cultural politics, so too did challenges to its status as a climate icon: lobby group the Global Warming Policy Foundation produced reports challenging polar bear-climate science links (for example, 'Twenty good reasons not to worry about polar bears', Crockford 2015).

Polar bears mean climate change

The parodic potential of polar bear imagery in cartooning began around 2008, but proliferated around the time of COP15 in Copenhagen, and the hacking of more than 1,000 emails from the University of East Anglia's

Climatic Research Unit (CRU). Since 1982, CRU scientists have been a key part of the climate change story, as they established a global temperature record of surface area temperatures from the 1850s to the present day. This temperature record has become a fundamental touchstone of climate science. When the emails and documents were stolen, they were then 'crudely misrepresented by climate deniers' (UEA 2021: lead paragraph). This international scandal, which became known as 'Climategate', rocked the COP15 negotiations; and longer term, resulted in a significant loss of trust in scientists and effect on people's beliefs in climate change (Leiserowitz et al 2013; see also Nature 2010). Data was cherry-picked to present the scientists at the centre of the scandal as frauds, and to present the CRU temperature record – and therefore climate science as a whole – as an elaborate and dangerous hoax (UEA 2021). For the scientists at the centre of the incident, there were huge personal repercussions as they received a barrage of hate mail and death threats. Phil Jones, the scientist at the centre of the incident, contemplated suicide several times (BBC 2010). Note that eight subsequent independent enquiries found no evidence of fraud or scientific misconduct (for example, Science and Technology Committee 2010). The incident has since been dramatised in a BBC film (BBC 2021).

I undertook my PhD at the University of East Anglia, and my first academic job after graduating from my PhD was a teaching role in CRU. To me, CRU was a group of leading climate scientists, but also a friendly and close-knit community of academic staff, PhD students and support staff who often downed tools mid-morning to tackle the cryptic crossword together over a coffee. I had only just moved to a new role at the University of Melbourne by the time of COP15 and the email hacking incident. As the incident unfolded in both international media and through conversations with friends and colleagues back at UEA, an ex-CRU colleague posted me a clipping of a cartoon from UK newspaper *The Telegraph* (Figure 3.2), in gallows humour: I, alongside past and present CRU staff and students, had received hate mail due to the hacking (although the abuse to staff at the centre of the claims was very much worse), so I was worried for the welfare of colleagues back in Norwich. Unlike most other CRU staff, I had also just published a paper using an expert elicitation approach to assess uncertainties in future polar bear populations under climate change (O'Neill 2008), so had recent exposure to the deep politicisation of polar bear science.

That polar bear cartoon clipping, a little yellow and crumpled round the edges 15 or so years later, has travelled back to the UK and remained stuck to my office wall ever since; a visual capturing some of my early experiences of science-society interactions around climate change. The cartoon depicts a sheepish-looking fluffy polar bear leaning against an iceberg and typing on a laptop as snow falls. It is captioned 'Secret email: Don't mention this,

Figure 3.2: Cartoon parody of the CRU 'Climategate' email hacking by Matt Pritchett, published in the UK *Telegraph* newspaper in 2009

Source: Author photograph of original newspaper cartoon clipping. © Matt/Telegraph Media Group Holdings Limited 2009

but its bloody freezing here'. The cartoon invokes themes prevalent in news reporting at the time, of scientific deceit and disrepute around the science of anthropogenic global warming.

By 2009, polar bear images seemed to be everywhere – I also encountered them at the 'Walk against warming' protest march in Melbourne, held as COP15 and the email hacking unfolded in Europe; where campaigners

Figure 3.3: Protester dressed in a polar bear costume. Walk Against Warming protest march, Melbourne, 12 December 2009

Source: Author's photo

polar bear costumes stood tall in the crowd and caused many to stop and take their own photographs (Figure 3.3).

As the 2010s began, polar bears had become firmly entrenched as a visual icon of climate change. Indeed, they had become irreducible from climate change – to see a polar bear image was to 'see' climate change (O'Neill 2022). This is evident in a polar bear image controversy from the scientific journal *Science*. In 2010, more than 250 scientists wrote a letter to the journal, titled 'Climate change and the integrity of science' (Gleick et al 2010). Its intention was to assert support for scientists who had experienced political assaults, and to underline the integrity of the scientific method, and in particular, of climate science. An editorial decision was made to accompany the article with a large image of a polar bear standing on a small, flat ice platform in a vast ocean (Figure 3.4). This was not a surprising decision, given the implicit association by that point in time between polar bears and climate change. Indeed, the image is entirely consistent with the visual cues of polar bear-climate change imagery (for example, Figure 3.1). However, this particular bear image was a composite image. Climate sceptic bloggers made much of the Photoshopped nature of the image, and the juxtaposition of this 'fake' and untruthful image against a letter about scientific integrity. The image was swiftly changed (to another polar bear image licensed through National

Figure 3.4: Composite image of a polar bear on a small ice floe surrounded by sea, as featured in the journal *Science* in 2010

Source: Coldimages, iStock

Geographic, an organisation known for not allowing any image alteration, Goldberg 2016) but the scientists' intentions had been both illustrated and undermined by the journal editors' image choice.

As polar bears and climate change have become ever more entangled in popular culture, the role that images are assumed to offer in terms of truth-telling have become increasingly evident. In 2017, *National Geographic* photographer Cristina Mittermeier was on a scouting trip in the Canadian Arctic for SeaLegacy, an organisation she co-founded which aims to use visuals to tell powerful stories and motivate action for ocean conservation. While there, she took photographs and a video of an emaciated polar bear, which the SeaLegacy team posted on Instagram. Its caption began: 'This is what starvation looks like'. Later on in the lengthy caption, it urged action on climate change (Mittermeier 2018: np). National Geographic reposted it with the caption: 'this is what climate change looks like'. The images and video quickly went viral, with an estimated 2.5 billion people viewing the imagery (Mittermeier 2018); and being named one of the top ten photographs of the year in *TIME* magazine (Katz 2017). While the video became National Geographic's most viewed ever, it also drew controversy. Some questioned the evidence for this individual polar bear's poor condition, and the link to global climate change. *National Geographic* later published an article with an apology, stating they 'went too far in drawing a definitive connection between climate change and a particular starving polar bear' (Editor's Note, in Mittermeier 2018).

The exemplar polar bear images this chapter has discussed have tracked the evolution of polar bear images over time. Polar bear images began as straightforward visual representations of a scientific story about declining sea ice and its knock-on impacts on polar bear habitat. They then become embroiled in social, cultural and political turmoil as climate change as an issue gained currency. Their status now is inseparable from climate change – they have become a visual metonym for climate change. As in Chapter 2, where images like that of Miriam Bulivono walking alone through floodwater come to move through the globalised media ecosystem to represent a much broader set of unspoken assumptions about climate vulnerability, so images of polar bears have become inseparable from wider assumptions about climate change politics (O'Neill 2022). The evolution of polar bear images in climate communication well demonstrates how the perceived truth, or provenance, of an image can entirely shape how it is interpreted and read by audiences, the power that image has, and the work that image might do in the world.

Flooding

Pictures of flooding are another common visual genre used to illustrate climate change (Smith and Joffe 2009, O'Neill et al 2013, Rebich-Hespanha et al 2015). As the climate warms, it will intensify the water cycle and therefore the amount and intensity of rain which will fall during wet events. However, there is a complex relationship between rainfall and flooding; with the type of river basin, the surface landscape, how wet the ground is before the rainfall event, and the extent and duration of rainfall, all influencing the likelihood of flooding. In sum, the severity of flooding is expected to worsen in some places, but could become rarer in other regions (IPCC 2021, see FAQ 8.2).

There is rather limited evidence about how people respond to flooding images in a climate change context; with studies focussed on the Global North. This literature suggests that images of floods can perhaps grab people's attention and be easily understood as being linked to climate change, but can also make people feel unable or unwilling to act. For example, in our Q-method study with US, UK and Australian participants, an aerial view of a large flood strongly invoked a sense of the importance of climate change, but also made people feel powerlessness about being able to take action on climate change (O'Neill et al 2013). These results were echoed in a similar study, using the same method, carried out in Austria, Germany and Switzerland (Metag et al 2016). A German-, UK- and US-based study also found an image of a man outside his flooded home resonated with participants, and was one of the easiest images to understand (Corner, Webster and Teriete 2015). However, an experimental survey in the US based on a mocked-up news story design found little evidence that seeing an image of flooding had any effect, either on people's sense of issue importance, or their sense of self-efficacy (Hart and Feldman 2016). And similarly, in a UK study examining visual communication of climate and health risks, flood imagery resulted in the lowest threat score, with poor air quality, for example, being viewed as much more threatening (McLoughlin 2020). The link made by participants (or the lack of it) between flooding and climate change is also important; Hart et al (2023) found that while a vivid photograph of a flooded house increased participants' perception of flood risk, it did not impact on their perceived threat of climate change. These studies deserve follow up, to understand more fully the impact of flood images on people's engagement; and to understand image engagement beyond the limited Global North contexts studied to date.

Flooding means suffering

Flooding images are particularly common when communicating about climate impacts, as Nerlich and Jaspal (2014) found in their analysis of news reporting the launch of the IPCC Special Report on Managing the Risks of Extreme Events and Disasters to Advance Climate Change Adaptation (also known as the SREX report). They analysed an online, global sample of English-speaking news reports covering the IPCC SREX report release. Nerlich and Jaspal found that flooding was the most common type of extreme event depicted in media imagery reported the SREX. Within the flood imagery corpus, they found two distinct groups. One group of images portrayed people and places where the context (such as people's clothing, or the architecture and urban design), suggested a location in the Global North. A second group of flood images depicted flooding in the Global South. This discussion on flood images now follows the overarching structure of the Nerlich and Jaspal study to first explore Global North flood imagery, followed by flood imagery of the Global South.

In the Global North flood images of the Nerlich and Jaspal 2014 study, people were inconvenienced or hampered, sometimes desolate, due to the flooding they were experiencing (images were typical of that shown in Figure 3.5). People's facial expressions were sombre, and clearly visible in the images. People were depicted as victims in need of assistance. Images focussed on the terrible and immediate aftermath of flooding and the human suffering that resulted. This finding echoes an earlier study of UK newspapers' climate imagery, which found flooding images showing scenes of disaster and chaos, such as cars stranded on flooded streets, or a person with a sombre expression walking in deep flood waters along a city street (Smith and Joffe 2009).

Figure 3.5: Typical flood rescue image

Note: Original caption: Morpeth floods, September 2008
Source: John Dal, CC BY-SA 2.0

Some have tried to challenge the dominant visuals of flooding as depicting only people struggling, often in the floodwater, in the period immediately following the flood event. Photographer David Maurice Smith lives in the Northern Rivers region of New South Wales, Australia. In 2022, the region was hit by two devastating floods a month or so apart. The floods were the worst on record, peaking two metres higher than the previous highest flood (BOM 2022). Maurice Smith at first found himself unmotivated to photograph the event, as it was already being featured heavily in news coverage in typical ways. Instead, he volunteered to help in the substantial clean-up effort. However, as coverage receded, he was galvanised to start to photograph the communities impacted by the flood, to try and document their lives as they began the long process of recovery. His work was motivated by a desire to try and draw attention to how while the floods themselves were catastrophic, the longer-term impacts were just as serious and worthy of attention (something also noted in the academic literature, for example Zhong et al 2018) – yet the media focus and visuals arising from the floods were entirely focussed on the short duration of the flood event itself:

> We were being bombarded, initially, with visuals. The flood had blown the doors off their expectations [of the flood level], it was beyond anyone's comprehension of what could happen. These pictures

showed the climate 'event'. But you hit a tipping point, with visuals. It creates a massive oversaturation and you have to look away. We were flooded with images. The human side, that's what's easily forgotten. (David Maurice Smith, photographer, NSW, Australia, personal communication, 22 April 2024)

Maurice Smith compiled a large collection of photographs of the flooded town of Lismore and its residents. His Instagram page highlights six Lismore photographs in a post about the project (one of the series is shown in Figure 3.6). The photographs are all portraits, focussing on different people's experiences of the floods. There is an 85-year-old man with his small dog standing between temporary housing portacabins; an indigenous Bundjalung elder finally sitting back in her home after it went underwater in the floods; a mother and daughter cross-legged on the floor in their empty home, still unsure of what would happen to their house more than a year on. In this compelling series of flood photography, it is notable that just one of the six images depicts flood water; in the other five, the flood waters are a spectre haunting through their absence, referenced more subtly via homes empty of furniture, flood-damaged walls and temporary housing. The stillness of the portraits visually evokes the frustrating waiting and uncertainty faced by people who have experienced a flood event.

Figure 3.6: Photograph from David Maurice Smith's Lismore flooding series

Note: Original caption: 'Ella Buckland and her daughter Myla in Lismore. The family lost everything when their home flooded.'

Source: © David Maurice Smith. Reproduced with permission

The Guardian led the news story about the floods with a large close-up from one of Maurice Smith's series. Notably, the image chosen was the only photograph which featured flood water, perhaps indicating that newsroom expectations of what a flood image 'should' contain hold firm. It shows a women holding a child on her hip at the edge of the flood (Figure 3.6; see Hinchliffe 2022). Houses, trees and power cables are visible at the edges of the image, and are reflected in the floodwaters. The woman looks directly at the camera, with a serious expression on her face. The girl cuddles into her shoulder, turned slightly towards the camera, with a neutral facial expression. The caption names them as Ella Buckland and her eight-year-old daughter, Myla.

The image is strongly linked to the text narrative in the Guardian piece (Hinchliffe 2022). The Bucklands' experience of the floods leads the introduction to the news article, and their narrative is referred to again in drawing the article to a conclusion. I found the Bucklands' photograph used in just one other news article: the second time, their photograph is used to accompany a *Guardian* Opinion piece authored by Ella herself, a year on from the Lismore floods: 'Before the floods I thought climate change wasn't my problem. Now, I'm not waiting for someone else to fix it' (Buckland 2023). It links directly back to the earlier news story. To date, the Bucklands' photograph has not been used as an image to illustrate any other *Guardian* climate news story, or elsewhere; it remains linked only to the reporting of the Lismore floods, and to the Bucklands' experiences in particular. This is partly a question of licensing and searchability: Maurice Smith has not listed these images on a stock image site (unlike the photograph of Miriam Bulivono in floodwater in Fiji, Figure 2.5) so they are unlikely perhaps to be picked up by news organisations searching for images of flooding and climate change. It is also notable, though, that Maurice Smith's other Lismore flood series images – which did not feature flood water directly – were not used in the Guardian's news reporting.

The Bucklands' photograph is reminiscent of Dorothea Lange's iconic 'migrant mother' image, which became emblematic of the US Depression, poverty and injustice (Lange 1936). Along with many others since, it draw on iconography of the Madonna and Child. This gendered iconography transcends geographical boundaries; its repeated visual cues common to conflict, war, development and aid photography (Banks 2001, Briggs 2003, Campbell 2010) and now into extreme weather and climate change photography. Such images are often taken outside, without the visible safety and stability provided by a home environment (Briggs 2003, Campbell 2010), and feature women and children but without picturing supportive others (particularly men). In contrast, the other five photographs from Maurice Smith's Lismore series do not fit within the migrant mother iconography, featuring, variously, visual themes of home (from a home that hangs in limbo between flood and repossession, to emergency housing pods) and depicting men (an 85-year-old man walking his dog; a middle-aged man alone in his flood-damaged home).

With the Bucklands' image, the threat of the flood is viscerally portrayed through the dark floodwater lapping at their feet – that is, the flood is not only connoted (symbolically portrayed), but clearly denoted (literally portrayed; Barthes 1977). As a former Deputy News Picture Editor at *The Times* described to me, there is a clear tipping point to visually capture in a newsworthy flood photograph – where water escapes from its usual location (such as a river) and becomes, in that moment, out-of-place (for example, overtopping a flood barrier, pouring through someone's front door, pulling a car downstream on a flooded highway) (Alastair Johnstone, personal communication, 5 January 2024). This moment of displacement, and the chaos it causes, clearly satisfies a range of news values, such as drama and magnitude (Harcup and O'Neill 2016). The Bucklands' image also uses the protective hip hold of a mother hugging her child close, sombre faces and an outdoors location without visible others to connote vulnerability (see Briggs 2003). The Bucklands' image fits a cultural expectation of what a flooding image should contain; it is consistent with the Global North flood images described by Nerlich and Jaspal (2014).

This visual trope of flood photography (depicting only the immediate aftermath of a flood event, of women or children, of people outside and away from supportive environments and so on) is not limited to the Global North; similar visual cues are evident in flood photography from around the world. For example, in 2023, Bangladeshi visual journalist Zakir Hossain Chowdhury won the South Asia Press Photo of the year 2023 award with the image 'The mother who saved her child from the rainstorm' (see Asian Art Association 2023). In his photograph, a mother walks close besides her older son while carrying her baby on her hip, thigh-deep in floodwater, trying to shelter them all from torrential rain under her umbrella. The photograph's tones are muted, except for the bright red of the mother's sari and the large bundle her son carries on his shoulder. Separately in the background, two other small groups are seen also making their way through the deep flood, but homes or people who could reduce the danger the mother and children are exposed to (such as rescue personnel) are not in the frame.

Flooding means 'getting on with it'

Nerlich and Jaspal's (2014) second type of flood imagery depicted flooding in the Global South. Here, visuals portrayed resilient people 'getting on with it' (Nerlich and Jaspal 2014: 262), carrying on living their lives despite the extreme flooding they were experiencing. Nerlich and Jaspal argue that the pictures portray flooding as a rather mundane sense of routine, as just an everyday part of life. This sense is enhanced by images depicting people going about their lives, through floodwaters, smiling or with a cheerful demeanour. Nerlich and Jaspal (2014) argue that these images do not engender compassion for the people depicted, as the flood is not depicted as something out-of-the-ordinary or concerning.

This visual trope of 'getting on with it', and even a cheerful demeanour when faced with flooding, is also still in evidence. In 2023, the United Nations posted a tweet about the impacts of climate change (see UNGeneva 2023). The tweet text says: 'Over 2 M deaths and $4.3 trillion in economic losses; that's the impact of a half-century of extreme weather events. The most vulnerable communities bear the brunt of weather, climate & water-related hazards.' Clearly, the context and content of the message in the text is deeply serious and concerning. However, the tweet's image depicts two laughing children standing shoulder-deep in a deep, grey flood, in heavy rain and under dark skies, holding big, bright green leaves above their heads as if they were umbrellas. The flood recedes into the distance, with the wider landscape also showing a flooded powerline and trees.

There is no context given (neither the names of photographer, the children nor the location are provided in either the tweet or alt text for the image), but a reverse image search discovered this photograph was taken in a village near Chittagong, Bangladesh by photographer Muhammad Amdad Hossain (Picfair 2023). The image composite is a clear example of the 'getting on with it' (in the Global South) images that Nerlich and Jaspal (2014) identified. Despite the disconnected and problematic nature of the text and image, it remains on the UN Geneva X (formerly Twitter) account and online for a UN website story at the time of writing (see UNGeneva 2023). This disconnect between text and image is reminiscent of the heatwave imagery in Chapter 2, where text and image are pulling in contradictory narrative directions (see also DiFrancesco and Young 2011).

Implicit visual cues

Nerlich and Jaspal (2014) suggest a distinction between the 'getting on with it' imagery of flooding in the Global South, and images portraying flooding as 'suffering' in the Global North. While other studies of visual representations of flooding do not find this distinction (for example, Ali 2014, Yu and Chen 2021), find many images of suffering in both Pakistani and Chinese news coverage of flooding, respectively), what is clear is that there are intersectional issues of stereotyping gender and race in the visual representations of flooding.

Visual representations of people experiencing floods are highly gendered. Visuals depicting men often dominate coverage of floods; while women are often invisible. Over half of all images in news coverage of a devastating flood in both the US (Vevea et al 2011) and Pakistan (Ali and Mahmood 2013) depicted only men. The US study found less than 10 per cent of images depicted just women (Vevea et al 2011). Men and women are shown playing distinctly different roles in much visual coverage of flood events, too. In a study of the coverage of the 2010 Pakistan floods, Ali (2014) found women were shown as helpless victims, who were relying on the strength and agency of men in order to survive. Similarly, in US flood coverage, men were shown as strong, capable heroes (Vevea et al 2011). Emotional portrayals can be gendered, too: in Pakistan, women were pictured showing inconsolable grief in the face of overwhelming loss, but the expressions of men were neutral and only concerned with the rescue and rehabilitation effort (Ali 2014). An important body of literature to mention here is that on the cultural barriers to climate adaptation, and the relationship between flooding and gender in climate adaptation in particular: for example, Chowdhury et al (1993; see also Jones and Boyd 2011) documented how local institutional restrictions in parts of south Asia prevented women from learning to swim, obliged women to wear clothing that inhibited swimming and constrained women's access to emergency warnings and cyclone shelters. These socio-cultural barriers are reflected in these depictions of women as helpless victims in flood imagery; and they also reinforce, repeat and recirculate these power relations.

Racial stereotyping is also evident in flood visuals. Kahle, Yu and Whiteside (2007) studied the news visuals associated with Hurricane Katrina in the US. They found news visuals consistently depicted Anglos in the role of helper, and African-Americans as helpless. They describe an overwhelming visual aesthetic of white military and social service personnel 'saving' African-American 'refugees'. While Anglos were pictured preparing for the storm, there was little focus on the preparations African-Americans made. Although low in overall number, they note high audience attention to images depicting African-Americans as looters, compared to Anglos standing guard against looting. Intersectional stereotyping is also evident in flood imagery. For example, in coverage of the flooding resulting from Hurricane Harvey, news images predominantly portrayed white men as heroes, saviours and caretakers; while Latinx women were depicted as powerless, displaced and distressed (Figueroa 2022).

These sexist and racist portrayals are deeply problematic. Climate impacts will be felt unequally; and especially by those already marginalised in society such as people from minoritised ethnic groups, women, children and people living in poverty (IPCC 2022). It is important that people's experiences are visualised in ways which are inclusive and representative. Stereotyped imagery influences how people perceive the people pictured, and impacts support for associated policies (Tukachinsky, Mastro and King 2011). For example, women are not only often at the frontline of climate impacts, but play a fundamental role in ensuring sustainable adaptation (Nellemann et al 2011), including in relief and management of flood events (Azad and Pritchard 2023). 'Damsel in distress' portrayals (Figueroa 2022: 455) ignore this important role and, instead, reinforce patriarchal norms.

Representations which depict people of African-American heritage as people who have not adequately prepared for a disaster, who are helpless and unable to help themselves and who are looters, portray African-Americans as irresponsible, suffering because of their own controllable actions and as dangerous and violent. It is notable that, where data is available, these stereotyped images are shown as the social construction via photojournalists, editors and wider society that they are. For example, Vevea et al (2011) discuss how there were an estimated equal number of women and men in flood relief efforts, not the 7:1 male:female ratio that eventuated in visual news depictions. Kahle, Yu and Whiteside (2007) show that Anglos were more likely to be pictured in news visuals despite the racial distribution of New Orleans City as predominantly African-America. And Ali (2014) contrasts the visual coverage of men crying in agony in live media coverage of the Pakistan floods, compared to the circumspect and neutral expressions of the men that eventuated in print news media visuals.

Where to next, for climate impact images?

This chapter has explored images used to portray the potential impacts of climate change. It has shown how the common climate visual tropes of polar bear images, and flood images, lean on notions of distant or 'othered' people, places and species; who exist without agency, and who require rescue by an incoming 'hero'. These visuals can act to distract attention from addressing the root causes of climate change. As Bennet, Foley and Krebs (2016: 24) discuss in the context of problematic stereotyping in famine photography, 'in the claimed act of raising awareness of the suffering malnutrition causes, the causes of famine – those things that ought to be addressed in order to prevent, alleviate or end famine – are overlooked or ignored'. In flood images, the white, male rescuer is situated as a hero, rather than (or as well as) someone with the power and agency to address the root causes of inequality acting upon minoritised people. Likewise, polar bear visuals: in situating climate change as distant in time and space, concerned with charismatic megafauna and not human (in)action, such images can be disconnected almost entirely from everyday political realities for addressing climate change.

4

Energy: Smokestacks and Wind Turbines

Energy, including where it comes from, how it travels and how we use it, is a fundamental part of climate mitigation and a key component of any discussion on climate change. Part of this discussion includes engaging with the aesthetic qualities of energy technology infrastructure and understanding how technology visuals inform discussions and negotiations around energy transitions. However, images of energy are seldom pictured alongside climate change news (O'Neill 2013, Rebich-Hespanha and Rice 2016). Where energy imagery is featured, it often elucidates energy as pollution, through the common visual trope of industrial chimneys belching out clouds of smoke or steam. When energy alternatives are pictured, their visual framing is fairly narrow; with the tall, sleek wind turbine featuring as a common visual motif of sustainable energy (León and Erviti 2013, Rebich-Hespanha and Rice 2016, Wessler et al 2016, O'Neill 2020). These visual tropes of climate and energy shape and reflect how we think and feel about climate change.

In the 2000s, a research agenda begun to emerge which sought to understand the values and symbolic meanings related to place and technology as a previously overlooked but essential part of energy transitions (McLachlan 2009, Devine-Wright 2014); with aesthetic qualities (particularly around whether a technology was perceived as in or out of place) a key part of public acceptance and controversy (Wiersma and Devine-Wright 2014, Gommeh, Dijstelbloem and Metze 2021). Devine-Wright and Devine-Wright (2009: 368), for example, discuss contrasting visual interpretation of electricity pylons in UK newspaper articles: from 'huge, ugly, monstrous' impositions on the landscape, to 'impressive, striking and dramatic' features. Others have sought to centralise the visual as a format with which to imagine, elucidate and negotiate energy futures; from approaches as diverse as photo-elicitation methodology for understanding renewable energy transitions (Devine-Wright and Wiersma 2021), to using thermal imaging technology in conversation with householders about energy conservation (Goodhew et al 2015) to communicating to policymakers about new forms of energy system via comic strip cartoons (Flockemann 2023).

This chapter examines two visualisations linking energy and climate, that of smokestacks and wind turbine imagery; both chosen for their status as common and iconic visual motifs. Immediately, it is notable that both of these visuals depict supply-side aspects of the energy debate; that is, they refer to how electricity or heat is generated, but not how energy is used by people, or how energy debates are made identifiable in everyday life; themes that the conclusion circles back to. In this chapter, there is a particular focus on the visual cues of perspective, hue, saturation, brightness and so on. Through the wind turbine and smokestacks examples, I demonstrate how narrow our conceptualisation of energy currently is; and how visual cues are used in predictable ways across energy imagery to communicate about climate risk. The chapter concludes by elucidating alternative ways to picture our energy futures.

Smokestacks

Smokestacks imagery has long been a visual cue for industrial pollution. For example, industrial chimneys belching out black smoke were featured in a Tony Auth cartoon in in the *Philadelphia Enquirer* in 1979, soon after the Three Mile Island nuclear accident. It featured US president Jimmy Carter leaning down to a flower to comment 'we're all going to have to make sacrifices … especially you, Mother Earth' alongside belching smokestacks (Dunaway 2015). Several decades later, it may seem that smokestacks have morphed into a ubiquitous visual for climate change. That may be the case now; but it has not always been so. I found that smokestacks imagery was rare in US and UK newspaper reporting on climate change before the mid-2000s. Smokestacks imagery was used just a few times, and only in US newspapers, between 2001 and 2005 (see Figure 2 in O'Neill 2020).

However, in 2006, Al Gore's film *An Inconvenient Truth* was released. The movie's main publicity poster, its DVD cover and other publicity materials all featured a stylised image (see Guggenheim 2006): seven chimney smokestacks silhouetted against an ominously cloudy and dark sky, with two smokestacks prominently emitting white smoke. The smoke curves as it rises upwards and forms into a feature resembling the clouds around a hurricane's eye. Through this image, the film made a somewhat contentious (at that time) scientific link between burning fossil fuels, anthropogenic climate change and the increasing risk of hurricanes. From 2006, US newspapers (the *Wall Street Journal* and *New York Times*) and UK newspapers (*The Telegraph* and the *Daily Mail*) started to use smokestacks as a small but regular part of visual coverage for climate change news (O'Neill 2020).

Figure 4.1: Typical smokestacks image

Source: Janusz Walczak, Pixabay

Typical climate smokestacks photography used by the news media (for example, Figure 4.1) carries strikingly similar visual cues. Ex-press photographer and now journalism scholar Helen Caple (Caple 2013) suggests that it is productive to analyse photography discursively both by taking into account image content (what is depicted) but also by considering what she terms 'camera technique', encapsulating how a photograph has been constructed; taking into account composition, perspective and technical considerations. In smokestacks photography, there are the strong verticals of the chimneys; either slim, tall chimneys, or thick, bulging, pale grey concrete chimneys. The chimneys may be either carrying potential polluting gases high into the air to disperse any dangerous pollutants over a wider area to minimise their impact; or be cooling towers emitting steam as a by-product of nuclear- or fossil-fuelled energy production. Whatever their output, chimneys emitting large clouds are used as a visual cue for industrial pollution and via this, to anthropogenic climate change (despite the difference in the chimney outputs: nuclear power plant chimneys emit only steam, and nuclear has a minimal carbon footprint (LSE 2022), unlike fossil fuels). However, the visual cue of spilling smoke also started to be subverted so the images are read in comedic or parodic ways, for example in the UK's *Daily Mail* in 2006, to play on the images' portrayal of fear and doom in an article titled 'Why the doom mongers have got it so wrong'; and using the smoke as a visual metaphor for exaggeration and hyperbole such as in an article from the Telegraph titled 'Hot issue or hot air?' in 2007 (see O'Neill 2020).

Smoke also provides aesthetic qualities of drama and movement to the image, especially where the wind blows the smoke sideways or into interesting shapes. Another similarity is the camera angle used to photograph the chimneys: often a straight-on shot, perhaps with a telephoto lens, to capture the sky and the smoke, but eliminating the ground from the image. When land does make it into shot, it may be covered in mist, or dark in silhouette. People are almost entirely absent from these visuals. The effect of this is to untether smokestacks images from most geographical landmarks and to turn them into generic markers of industrial pollution, disconnected from everyday human landscapes and experience.

Media and communications scholar Giorgia Aiello and her colleagues have written compellingly about 'generic visuals': the way in which images, particularly stock photos, come to have standardised appearances and formats which circulate and proliferate through the news media due to changes in the media ecosystem (Aiello et al 2022). Image agencies are therefore powerful sites of visual meaning-making. Smokestacks imagery as used in climate communication is an example of this generic imagery. In being geographically unsituated and in lacking a connection to a particular place, smokestacks imagery stands in for (and is tagged for, by commercial image agencies) the generic issues of 'pollution' and 'climate change'. This movement towards generic, banal visuals sees images becoming more commercially valuable, with the same image used in different article and outlet types, across quite different geographical contexts: as they are both easily repeatable and widely recognisable (Frosh 2003, Hansen and Machin 2008, Aiello 2023). Indeed, smokestacks imagery appears to be a top-of-mind association for users searching the Getty Images database for climate-related images: a search for 'climate change' images, with the 'creative' and 'most popular' filter toggled on, gives seven smokestacks images within the first page of 46 image results (search performed 30 June 2023; with similar results when repeated 07 July and 22 August 2023). No other image type or visual trope had a greater share of the visual representations returned.

My colleagues Ranadheer Malla, Tristan Cann and Ned Westwood (all Computational Social Scientists) performed a search for 'smokestacks' imagery featuring on the Getty Images site (again toggling the 'creative' image type and 'most popular' settings). The results, in Figure 4.2, compellingly portrays the visual cues typical of smokestacks imagery described earlier (Figure 4.2; first 100 pages of search returns, n=5,988 images, collected 11 November 2023). All images returned by the search were plotted on a colour wheel. There is a cluster of blue images, most of which feature an azure sky background. There is also a substantial cluster of dark orange-red images, which feature dramatic sunsets (Figure 4.2, lower inset). There are fewer images in the green segment (Figure 4.2, upper inset).

Figure 4.2: Colour wheel of images arising from a Getty Images search for 'smokestacks', 11 November 2023

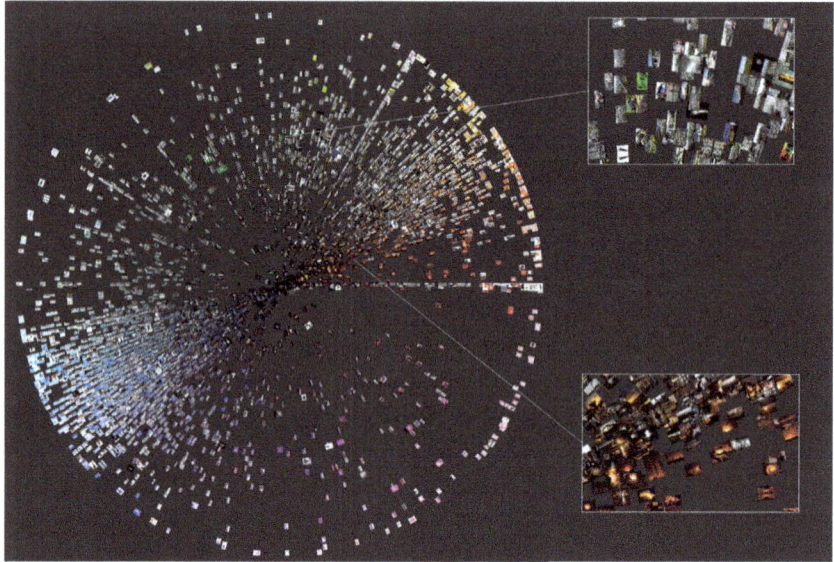

Notes: Images plotted on a colour wheel with median colour hue (y-axis) and brightness (x-axis). Plotted using ImagePlot. Figure created by Ned Westwood.
Source: Reproduced with permission

The colour hue of smokestacks imagery can be used as a visual cue indicating its emotional quality, or 'valence', in the context of, for example, a news media story about climate change (Lang et al 1990, Kress and Van Leewen 2002, Lehman et al 2019, O'Neill et al 2023: where positive valence may indicate something is fun or enjoyable; negative valence that the issue is something to be concerned or worried about, that it may be dangerous or risky). For example, I carried out a TinEye reverse image search, using a typical Getty Images smokestacks image which featured a bright-blue sky background. This blue-hued smokestacks image only accompanied news text which was calculated as neutral or positive in sentiment when analysed by a text analytics AI (Text2Data 2023). The headlines and lead paragraphs accompanying this blue sky visual discussed the need for greenhouse gas emissions targets but within the context of the UK being an international climate policy leader, or headlined good-news results of greenhouse gas emissions falling and fossil fuel use decreasing: including examples from CNN: 'The UK just went one week without coal power for the first time in 137 years' (Mezzofiore 2019); the *India Times* (Reuters 2020) 'UK greenhouse gas emissions fell 3.6 per cent in 2019'. At *The Independent*, the headline 'Climate Change Act must set "net zero" emissions target by 2020, experts say', had a lead stating: 'We have had responsible ministers from all parties who have pushed the act – an amazing amount of consensus compared to the US or Australia' (Gabbatiss 2018).

In contrast, a TinEye reverse image search for a Getty Images smokestacks image which instead featured a fiery orange background with three smokestacks pumping out smoke silhouetted in black against a glowing sun only accompanied news stories which were assessed as negative in sentiment. The news articles using this orange smokestacks image discussed climate change as a significant danger, risk and threat, for example: 'IPCC climate change report: summary of a wake-up call for the world' in *The Times* (Burgess 2021) and, on the BBC News website, 'COP27: Climate change threatening global health – report' (Hambly 2022). The red or orange hues in these images can evoke feelings of heat, as well as threat or danger (Schneider and Nocke 2018).

This is not to say that there is a conscious choice made by journalists or editors when selecting images and their hued backgrounds, but that cultural norms around colour use (see also Chapter 6 in discussion of climate science visuals) may play a role. For example, *The Independent* newspaper article featuring the blue smokestacks image was authored by climate journalist Josh Gabbatiss. When asked about any intention he may have had for the choice of image for this story, he commented: 'It was a busy newsroom environment at The Independent, so I don't recall putting a lot of thought into the mood I was trying to create with that image – but maybe I picked images that matched the mood subconsciously!' (Josh Gabbatiss, Science Correspondent (2017–2019), *The Independent*).

This section details how smokestacks rose up in terms of media prominence, to become deeply and historically embedded in the meaning-making of climate change; including through their use as parodic and clichéd representations. Smokestacks have become a synecdoche – a form of visual shorthand, used within a particular culture to immediately signify to the reader a particular set of ideas about climate change beyond the immediately represented content (O'Neill 2019). In other words, there is 'more than meets the eye' (O'Neill 2019: 1) to how these smokestack visuals are being produced, read and circulated. I now turn to another type of visual synecdoche associated with climate and energy debate, that of wind turbines.

Wind turbines

There was very little visual coverage of climate mitigation measures in the news media compared to other types of visual climate coverage such as images of climate impacts (O'Neill 2013). Of the very little imagery that does make it into the news media, wind turbines, a form of renewable energy technology, are the most common. Devine-Wright (2014) found wind energy the most common type of renewable energy featured in visuals in UK newspapers. Wind turbine visuals are rarely used before 2005, but appear more frequently from the mid-2000s onwards (see Figure 2 in O'Neill 2020). They have become a well-used symbol of hope within the environmental movement; see, for example, the repeated use of wind turbine iconography in The Climate Collection, a global gallery of digital illustrations of climate change, with an aim to focus on hope and solutions (Artists for Climate 2024).

While wind turbine imagery is used to accompany news stories directly about energy, and wind energy specifically, wind turbine images also start to be used in satirical and parodic ways to illustrate climate sceptical narratives from the mid-late 2000s onwards. For example, in 2008, the UK's *Daily Mail* used an image where the camera angle emphasised the size and movement of the turbine blades to accompany the article 'Greenwash!; A Cambridge don argues that we are being misled by green propaganda' (O'Neill 2020). This polysemic character (that is, the ability to carry several different meanings) is increasingly evident elsewhere; for example, Norwegian social anthropologist and journalist Anne Hege Simonsen explored the visual representations of wind turbines in a diversity of Norwegian media, at two time points (2018 and 2020). Although wind turbines were predominantly presented as 'future perfect', hopeful green icons in 2018, they were transitioning to also symbolise nature degradation and political arrogance by 2020 (Simonsen 2022). A typical wind turbine image (for example, Figure 4.3) depicts wind turbines as sleek, white, clean, majestic machines against a vivid blue sky. Wind turbines can be situated either on land, or at sea; with potentially different readings of the image depending on whether and how this geographical location is depicted, that is, whether the image is (sometimes quite literally) 'grounded' in socio-political context, such as showing protesters at wind turbine construction sites (for example, Simonsen 2022).

As discussed, image agencies are powerful sites of visual meaning-making. Again then, as for Figure 4.2, my colleagues Ranu, Tristan and Ned searched for 'wind turbine' images on Getty Images; again toggling the 'creative' image type, and 'most popular' image search functions. Figure 4.4 represents the results from the first 100 pages of search results (n=5,992 images; collected 11 November 2023) in a colour wheel. Like the smokestacks images, there is a cluster of deeply saturated, bright-blue images of azure skies. However, deep and bright green-hued images also feature (Figure 4.4, top insert). Most images in the Getty search show land-based wind turbines, though some also feature solar panels as well as wind turbines, or marine rather than terrestrial environments. It is notable that, unlike the smokestacks images, many of the wind turbine images depict landscapes recognised for their beauty, at least in the European imagination (van Zanten et al 2014, Caple and Bednarek 2016).

Figure 4.3: Typical wind turbine climate change image

Source: Boke9a, Pixabay

Another cluster of images feature orange and pink hues (Figure 4.4, lower inset), typical of the images taken during the 'golden hour' in photography (the hour after sunrise and before sunset), to enhance the light quality of an image. In contrast to the 'dangerous' or 'hot' red-orange hues in the smokestacks colour wheel (Figure 4.4), the orange-pinks of the wind turbine backgrounds construct a peaceful, ethereal or dreamy quality.

Compositional elements, perspectives offered and technical considerations are similar across the wind turbine images. In terms of composition, elements are often arranged similarly in the frame (Caple 2013). The photographs often draw on the 'griddiness' of the turbines, through the turbines' regular, grid-like placement in the landscape. These linear and grid qualities are drawn out to either show the turbines stretching endlessly to the horizon, or to fill the frame with regular, repeating pattern. The amount of sky featured in the photographs is also consistent, with many featuring the 'Rule of Thirds' (Caple 2013) in one third landscape, two thirds sky. And of these skies – they are uncluttered with other features, for example showing cloudless, azure skies, or beautiful sunset scenes.

Many of the wind turbine images use either a shutter speed which is fast enough to capture the turbine blades without showing their movement through blurring or trails; or are images which have been taken during still weather. This has the effect of removing a sense of motion or energy production from the viewer.

Photographers of the wind turbines images often make use of a long shot. This creates an impersonal distance between the viewer and the object (Kress

Figure 4.4: Colour wheel of images arising from a Getty Images search for 'wind turbines', 11 November 2023

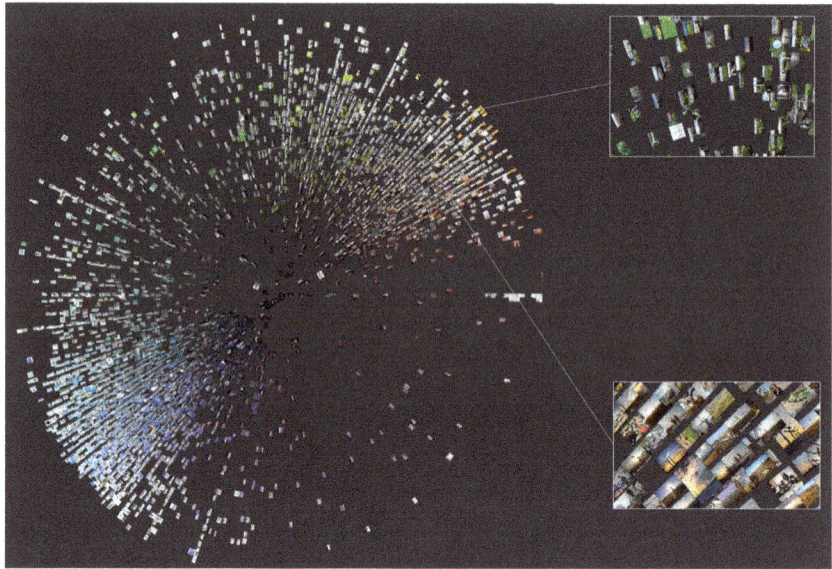

Notes: Images plotted on a colour wheel with median colour hue (y-axis) and brightness (x-axis). Plotted using ImagePlot.

Source: Figure created by Ned Westwood. Reproduced with permission

and van Leeuwen 2020). In the many low-angle and eye-level shots, the wind turbine technology is given symbolic power over the viewer. Unlike high angle shots, the technology is positioned out of reach and command of the viewer (Kress and van Leeuwen 2020: 139). In common with much landscape photography, many of these images capture as much as possible of the landscape, perhaps with a wide angle lens, increasing the epic quality of the images. Geographer Denis Cosgrove, writing about the Apollo 11 'blue marble' images, powerfully argues that these iconic images can be read through either a 'whole-earth' interpretation (the impressiveness, ingenuity and powerfulness of humans gaining mastery over nature) or through a 'one-world' interpretation (the imperative for society to protect the environment, given Earth's smallness in the vastness of Space; Cosgrove 1994). These opposing readings result in entirely different conceptualisations of the power of the image, and the work it performs in society. Similarly, then, these wind turbine images could be read in two very different ways: either a 'technological new dawn' reading (the impressive, powerful and redemptive promise of this renewable technology); or a 'looming monsters' reading (making the viewer feel small, insignificant and powerless in the face of a landscape-dominating technology).

Indeed, public engagement research has demonstrated that wind turbines have been construed, in the news media, as 'looming monsters' out of place in the countryside; and that the height of turbines is perceived as an unwelcome and unnatural intrusion on a valued 'natural' landscape. A 2011 article in the UK's Times newspaper reports of people's 'astonishment that anyone could be so mad, so vandalistic as to site these monstrosities in such a way as to interfere with the view' (cited in Batel and Devine-Wright 2021: 52). Wide angle lenses to capture the potential visual impact of wind energy technology of a landscape have also proved controversial, with both the UK news media (Preece 2012) and academic research (Takacs and Golden 2019, Simonsen 2022) indicating that wide angle shots can be construed as making wind turbines look smaller than they really are; an important consideration given that aesthetics can be an important factor in wind technology planning applications. In contrast, research examining co-occurrence of wind energy visuals on social media platform Instagram found a high proportion of wind energy images were associated with positive emotion words (Vespa et al 2022), consistent with a 'technological new dawn' reading of these images.

As with the smokestacks images, few of the Getty search results for wind turbine images feature people; a finding common to news media coverage (Devine Wright 2014, O'Neill 2020). Where people are pictured, they are alone or in pairs, and – by virtue of holding equipment, or being featured in precarious positions on the turbine shaft – look to be completing technical tasks associated with the turbine technology. These images give a more human scale to these climate-energy images, again, in contrast to the smokestacks images.

These results suggest some broad similarities in the Getty Images collection, in terms of wind turbine imagery. This is important, as image agencies in general, and Getty in particular, has been shown to dominate visual coverage of some climate topics in the news media (for example, Hayes and O'Neill 2021, Hayes and O'Neill 2024). But different visuals portrayals of wind turbines are evident in the news media, and perhaps especially so as wind turbines gain traction (and opposition) as a proposed option for addressing the energy transition. Simonsen's study of Norwegian media wind turbine imagery suggests that commercial image sources (both industry actors and image agencies) tend to portray wind turbines in similar ways, as industrial icons. But, in contrast, she found that photojournalistic visuals related more specifically to identifiable contexts.

Putting people in the (energy) picture

As the introduction set out, it is notable how both of these common energy and climate visuals, smokestacks and wind turbines, depict only the supply-side aspects of the energy debate. Energy transitions and decarbonisation scholar Patrick Devine-Wright comments (personal communication, 12 July 2023):

> In my view, the prevalence of these common climate and energy images are problematic for socially embedded, fair energy transitions. By lacking people or a depiction of the demand-side aspects of the energy debate, they perpetuate a problematic, yet unfortunately conventional 'technocentric' approach to energy transitions in policy discourses and beyond.

Research with audiences in the US, Australia and Europe indicates that energy imagery can have particular effects on viewers. For example, images of smokestacks may be effective in bringing attention to the issue of climate change (that is, increasing a sense of issue salience) but they can also leave viewers feeling helpless or overwhelmed (that is, decreasing a sense of self-efficacy) (O'Neill and Nicholson-Cole 2009, O'Neill et al 2013, Rebich-Hespanha et al 2015). It is important though to note the need for further research here, as Hart and Feldman (2016) did not find the same effects replicated in their US-based survey study. In contrast, images of renewable energy appear to be particularly effective at invoking positively valenced emotions (Rebich-Hespanha 2011), and increasing a sense of being able to take action on climate change (O'Neill et al 2013, Hart and Feldman 2016, Metag et al 2016).

Images which show a more relatable sense of energy futures can be important for motivating feelings of self-efficacy, particularly when they show 'real people' (O'Neill et al 2013; Corner, Webster and Teriete 2015: 5). When energy is mostly represented through images of technology, at a grand scale and removed from a sense of everyday life, audiences may struggle to connect to other highly relevant issues around energy use and climate change, such as changing consumption habits or reducing energy demand (Remillard 2011, Rebich-Hespanha and Rice 2016). To date though, images which visualise climate and energy as something closer to home are rare, both on television and in newspaper coverage (O'Neill 2013, León and Erviti 2013). Indeed, journalists are frustrated at the lack of more relatable, person-centred climate-energy imagery, which they have also found lacking in commercial image collections. For example, Geographer Sylvia Hayes, when undertaking a newsroom ethnography at the UK-based climate journalism organisation *Carbon Brief*, reported the team discussing the substantial difficulty in illustrating a piece about energy transitions: 'when we've tried to find [images of heat pumps] before, they've been really boring' (Carbon Brief journalist interviewed in 2021; cited in Hayes 2024).

The Carbon Brief journalists discussed how heat pump images currently lacked the aesthetic qualities of, for example, wind turbine images and their

Figure 4.5: Typical image featuring a heat pump

Source: HarmvdB, Pixabay

beautiful skies, landscapes and colour hues (see typical image, Figure 4.5). Also, the Carbon Brief news team favoured images with people in, but found most heat pump images did not feature people, and when they did, they only depicted the heat pump installer (Hayes 2024). This seems to be a frustration felt more widely among journalists; with similar reports of newsroom debates around how best to illustrate a story around heat pumps – and the difficulty in tracking down compelling heat pump images – arising in my conversations with both US and UK journalists.

An additional issue with trying to bring people into everyday energy imagery is in avoiding clichéd or stereotyped visuals. In German- and UK-based survey and focus group research, Corner, Webster and Teriet (2015) investigated how viewers engaged with an energy efficiency image provided by Transition Belsize. The image depicted a grinning dad and son carrying out DIY by fitting draught-proofing around their kitchen door frame, while the mum and young child leaned, laughing, into the frame in the background (see image 24 in Corner, Webster and Teriet 2015: 13). Portraying similar visual cues in a commercial setting, the 'Family lifestyle travel wind energy' collection by Me 3645 Studio (Getty Images 2023) contains a selection of 15 similar photographs of a mum, dad and child laughing, hugging and high-fiving each other in front of wind turbines under a beautiful sunset. The caption for the collection reads: 'Parents take their children to visit and admire renewable energy windmills. The green technology industry produces fuel and energy from wind energy'. These contrived and staged energy futures images are likely to be read as clichéd, inauthentic and even ridiculous (Corner, Webster and Teriet 2015). Typical energy pictures also tend to reinforce unhelpful gendered norms, where men are mostly pictured acting on energy, despite the key role of women in the energy transition (European Parliament 2019).

A compelling alternative to trying to match (often unsatisfying) energy visuals to a climate and energy story is demonstrated via a move to solution-based journalism. As the Solutions Journalism Network mission statement explains, this entails a shift in news reporting (in general, not just news visuals) and 'focusing reporting on responses to problems and what we can learn from their successes and failures' (SJN 2023: np). Emerging research suggests that, in contrast to other forms of journalism, solutions journalism may allow journalists to effectively communicate about the scale of climate risks, but without decreasing support for climate action (Thier and Lin 2022). Neuroscientist Kris De Meyer reported a similar approach, 'action-based storytelling', in a climate-energy case study during a UK House of Lords evidence session investigating how to mobilise climate action (House of Lords 2022, Corner 2022). De Meyer noted how, in their action-based storytelling training, journalists were encouraged to turn from an issue-based story (in this case, around retrofitting homes with a heat pump, and whether the heat pump was good or bad) to instead focus on an action-based story (following the story of UK pensioner Rose Lewis navigating the Government green grants system, including fitting a heat pump). The visuals which accompanied the story therefore focussed on the pensioner in the story and the benefits the heat pump provided – Rose Lewis was picturing smiling in her cosy kitchen (Figure 4.6) – rather than the visuals featuring the 'boring' white box technology of the heat pump itself (as Figure 4.5). There is also climate justice angle here: using visuals that demonstrate the benefits of emerging energy to groups that stand to benefit the most could be a powerful engagement tool. The story was published in a regional newspaper (Atkins 2022) and was so successful it was then also picked up by ITV, a British television channel that broadcasts nationwide (House of Lords 2022).

Another alternative approach to imagining different energy futures through visuals is exemplified by the collaborative research of Science and Technology Studies (STS) scholar Ola Michalec and public engagement expert Joe Bourne. In their project 'Electric Feels', exhibited at the Victoria and Albert Museum in London, they invited four artists to respond to energy transformation research, focussing on the intersection between energy, digital technology and emotions (V&A 2023, Petras 2024). Electric Feels was concerned with the often invisible nature of energy production, delivery, management and maintenance; and so was a deliberate intervention and

Figure 4.6: An energy story created using a climate solutions journalism approach

Note: The story featured an image of Rose Lewis in her now-warm kitchen, rather than featuring the heat pump technology itself.
Source: Alex Leat/Local Storytelling Exchange

invitation to stop and question how we feel about energy. Ola Michalec and Joe Bourne prompted the artists to grapple with questions such as 'what is your vision for a sustainable future?' and 'how could your domestic routines and chores change as a result of future energy innovations?' (and through the artists' visuals, so too viewers grapple with these issues; V&A 2023 and personal communication, Ola Michalec, 5 May 2023). As Ola Michalec summarises:

> Digital innovations in the energy industry bring both exciting promises and a set of serious ethical concerns, which we must share with the members of the public. Over the past few years, it's been really challenging for the UK Government to open up a conversation about smart meters, heat pumps or energy tariffs, as the typical 'look' of those technologies – predominantly boring, grey devices – does not inspire engagement. We're hoping that by showing human negotiations, experimentation and the mess and energy transition, we'll be able to prompt new audiences to reflect on the role of technology in sustainable transitions.

One of the artistic responses is by digital illustrator Mary Flora Hart. In that image, Hart highlights a series of brightly coloured shapes and lines as the usually invisible energy is generated and used as it flows around a kitchen-dining area in a busy shared flat. Multiple smart energy devices glow as they challenge the presumed 'smartness' of the consumer, to represent both the possibilities and difficulties of a 'Home Control Room' in a climate-energy future vision.

Where to next, for climate and energy imagery?

Countries are increasingly implementing national policy directives to realise their climate mitigation ambitions, as part of the Paris Agreement commitments to reduce their greenhouse gas emissions (for example, the UK's Net Zero Strategy (BEIS 2021), and Pathways to Net Zero Greenhouse Gas Emissions by 2050 in the United States (US Department of State 2021)). Public debate around energy production, travel and use have become increasingly heated: in the UK, this is seen in the response to pressure group Insulate Britain (campaigning for the UK government to carry out a low energy and low carbon retrofit of UK homes by 2030; BBC 2021) and the intense political debate around the campaign to expand London's Ultra Low Emissions Zone (ULEZ) (Badshah 2023). More widely, energy debates have come to the fore in the stinging debate and even conspiracy theories surrounding the 15 Minute City concept. This has seen the urban planning concept conceived to promote sustainable and healthy communities – ensuring communities have access to essential facilities like schools, parks, shops and leisure with a 15 minutes of active travel (walking or cycling; and not needing to be dependent on a car, for example) – met with deep opposition (Stanford 2023).

The polysemic nature of images means that visuals play important roles in shaping public engagement with energy, in sometimes unanticipated ways. Strategic communication scholars Amber Krause and Erik Bucy (2018) show this in their study of how people engage very differently with the same set of images of that most contentious form of energy production, fracking; and how that can shape their perceptions of environmental risk. The wind energy visuals discussed are a further example of the importance of understanding polysemy for the climate-energy debate; where the same visual image of a wind turbine could be interpreted in diametrically opposing ways. It is clear that visuals of climate and energy – whether they depict where energy comes from, how it travels or how we use it – are already, and will increasingly become, a critical part of public debate.

This chapter has examined the images used to represent climate change and energy. It has discussed how, despite the central importance of energy debates to climate action, energy visuals are low in number in the news media and lean on a few aesthetically similar, generic images. While these images are easily recognisable, and may link to well-worn environmental discourses (of pollution – in the case of the dark, gloomy smokestacks plumes), they may also provide opposing polysemic readings (of 'technological new dawn' or 'looming monsters', depending on the interpretation of the towering wind turbines). Current energy images may fail to make much of a link between societal debate around energy futures and people's everyday lives. Calls for a move towards climate solutions journalism could be key, here, to reframe energy stories away from issue-based narratives and towards action-based stories; bringing with them substantial opportunities to open up the climate and energy visual discourse.

5

Science: Climate Stripes and Weather Maps

We have come to know about climate change through the work of science and scientists. Careful observation, measuring and modelling has alerted humanity to the transformation happening in the world's climates (Hulme 2009). This scientific work has spread widely, with some images becoming iconic in the climate debate. This chapter looks first to the 'Burning Embers' diagram, a historic climate science visual which has featured in the IPCC reports since 2001. It then discusses two climate science visuals which have gone beyond the science-policy interface and entered popular culture; the 'Climate Spiral', and the 'Climate Stripes'. The Climate Stripes began life as a visual aid for a scientist presenting at a literary festival, but quickly proliferated across everyday life: from the BBC news homepage to the cover of Greta Thunberg's book; to dresses, football kits and high fashion; racing car wraps to public infrastructure; even cushion covers and shower cubicle tiling. The Climate Spiral, meanwhile, is possibly the most-viewed climate science visual ever, as it was featured at the two-billion strong audience for the Rio Olympic games opening ceremony in 2016 (Hawkins et al 2019). The chapter finishes by bringing together a discussion of the IPCC's 2023 'Climate Generations' figure (itself partially a reworking of the Climate Stripes) and the visual affordances of the weather map; with a particular focus on the controversy both have engendered through use of the colour purple.

The attention on scientific imagery in this chapter facilitates an exploration of the role of images as truth-sayers and proof-makers for climate claims (Doyle 2011) – both in formal climate politics (particularly as negotiated through the UNFCCC and IPCC), but also in the broader cultural climate politics (Bulkeley, Paterson and Stripple 2016) of literary festivals, sporting mega-events, showcased in people's homes and even displayed on their bodies. Scientific visual representations of climate change are the 'showplace of science' ('Schauplatz der Wissenschaft'), where scientific knowledge becomes amalgamated into a visual form stimulating wider debate

(Heintz and Huber 2001: 34, cited in Schneider 2016). Although scientific visuals appear to account for a relatively low number of visuals in the news media (O'Neill et al 2013 study of US, Australian and US print media found scientific visuals accounted for around 6 per cent of all newspaper visuals) they do appear to be associated with particular framings of climate change. O'Neill et al (2015) found scientific visuals particularly associated with the 'settled science' climate news issue frame, which emphasises the science of climate change, and uses this evidence in a call for action.

Scientific images have been particularly important for the climate debate, partly as 'climate change' is a scientific construct that can never be visually represented (Hulme 2009), so scientific images have become a key way to represent and imagine unknown, potential futures (Schneider 2016). Scientific visuals are also important because climate change is a topic where so much of the focus has been centred at the science-policy interface, so visuals can act as important boundary objects working at and between these spaces (Mahony 2015). This chapter challenges the oft-assumed reading of scientific images as representations of an objective reality, and instead picks apart how even scientific images are highly normative statements about how the world works (Urry 1992, Hall 1997).

Burning Embers

The Burning Embers diagram (Figure 5.1) has a more than 20-year history as a boundary object at the climate science-policy interface. It was first conceived in 2001 as part of the IPCC's Third Assessment Report. It represented the substantial discussion around delineating what might construe 'dangerous anthropogenic interference with the climate system', a key part of the climate change legal framework of the United Nations Framework Convention on Climate Change (UNFCCC, Article 2, UN 1992). Defining what might be considered 'dangerous' is a value judgement, and therefore a task beyond the remit of the policy-relevant, but not policy-prescriptive, work of the IPCC (Dessai et al 2004). The figure was created with the intention to represent a review of the academic literature on climate impacts, and as a visual aid for decision-makers to then make their own value judgements of what might be considered 'dangerous' in terms of climate impacts (Mahony 2015). The following discussion focusses on the aesthetic choices of the figure, though note other criticisms have also been levelled at the diagram; such as its focus on the portrayal of additional risk posed by climate change, rather than portraying the change in total risk (which would include climate change as one of many changing factors including income growth and trade regimes; O'Neill 2023).

Human geographer Martin Mahony carried out a series of illuminating interviews with the IPCC scientists who created the Burning Embers,

Figure 5.1: The 'Reasons for Concern' (or 'Burning Embers') IPCC diagram, which featured in the IPCC Synthesis Report, Summary for Policymakers, Cambridge University Press

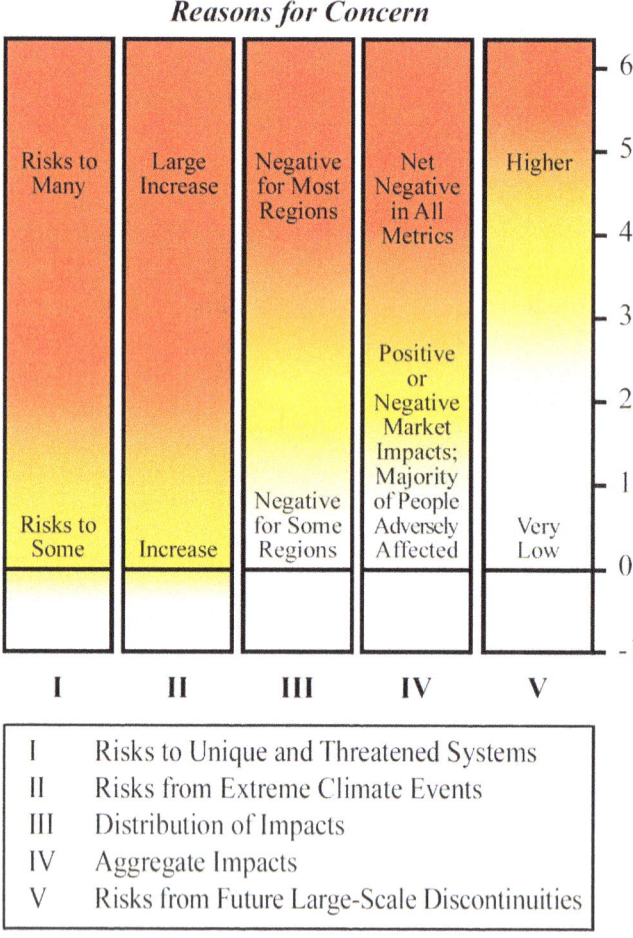

Notes: Original caption: 'Figure SPM-2: Reasons for concern about projected climate change impacts. The risks of adverse impacts from climate change increase with the magnitude of climate change. The left part of the figure displays the observed temperature increase relative to 1990 and the range of projected temperature increase after 1990 as estimated by Working Group I of the IPCC for scenarios from the Special Report on Emissions Scenarios. The right panel displays conceptualisations of five reasons for concern regarding climate change risks evolving through 2100. White indicates neutral or small negative or positive impacts or risks, yellow indicates negative impacts for some systems or low risks, and red means negative impacts or risks that are more widespread and/or greater in magnitude. The assessment of impacts or risks takes into account only the magnitude of change and not the rate of change. Global mean annual temperature change is used in the figure as a proxy for the magnitude of climate change, but projected impacts will be a function of, among other factors, the magnitude and rate of global and regional changes in mean climate, climate variability and extreme climate phenomena, social and economic conditions, and adaptation.'

Source: IPCC 2001b

to understand its genesis (Mahony 2015). He describes how this visual purposefully uses a number of visual cues to invite subjective judgements about how 'dangerous' might be defined. The figure makes use of a white, yellow, orange and red colour scheme to make the transition to danger; relying on the social convention and practice of red signalling 'stop' in a traffic light series. It is perhaps less clear what white represents: an early iteration of the figure used the traffic light's green as the starting colour, but was quickly changed:

> There were many incarnations of the whole figure and one was a different colour scheme with green, yellow, red. The green – a lot of people interpreted it as 'no risk' … which is a little bit of a different message. The colour scheme actually makes a very big, big difference in that diagram. (Interview with an environmental scientist; quoted in Mahony 2015)

It is worth noting that the UK Meteorological Office (Met Office) National Severe Weather Warnings risk matrix was also first pictured using green alongside amber and red, but they too removed the colour green (changing it to grey) with the same rationale – as green shading might indicate no risk (Helen Roberts, socio-meteorologist, Met Office; personal communication, 5 January 2024). An earlier version of the Burning Embers also tried out using a blue hue to indicate potential positive effects of anthropogenic warming, in the aggregate column (column IV, see Figure 5.1); as some studies had indicated potential economic benefits from low levels of warming. This was also changed as discussion among the scientists revealed a lack of widespread scientific literature supporting this visual positioning (Mahony 2015).

The Burning Embers figure makes use of a hue fade between each hue change. So, there is a hue gradient between orange into red, for example. This hazy shading, or 'visual cloudiness' (Mahony and Hulme 2012: 5), was designed to make clear that a value judgement (using one's own interpretation and values) was required by the reader: 'The chart is sort of skilfully blurred to make sure you don't have an on–off switch. [We] deliberately didn't want that, because we weren't able to say "1.9 degrees good, 2.1 degrees bad". It doesn't work that way, so we blurred it' (Interview with economist A; quoted in Mahony 2015).

Although the economist in this quote describes this as a skilful blurring, intentionally designed to force decision-makers to confront the scientific evidence on potential climate impacts and make clear the value judgement required to assess what was a 'dangerous' level of climate change, it is notable that the blurring became a key part of the debate surrounding the visual: in terms of what was, and was not, acceptable as a visual cue in a scientific image (Mahony 2015).

The Burning Embers featured prominently in the IPCC's Third Assessment Report, featuring in the Summary for Policymakers (IPCC 2001a, see Figure 5.1). It subsequently travelled into the broader climate change debate, and became subject to fiercely competing claims about how the figure could be read in terms of representing science, uncertainty and policy (Mahony 2015). Although an updated embers figure was put forward for the IPCC's next edition, the Fourth Assessment Report, a fraught plenary session featuring officials in opposition (representing the United States, China, Russia, and Saudi Arabia) versus those keen to include it (from Europe, Canada, New Zealand and Small Island States) eventually saw the figure excluded from the report (Steve Schneider, in Mahony 2015). The updated Burning Embers figure was eventually published as a separate paper in a scientific journal (Smith et al 2009). The claims and ensuing controversy of the figure have continued; though it did reappear for both the IPCC's Fifth Assessment Report (this time, with 'very high risk' represented by the colour purple; IPCC 2014) and into the Sixth Assessment Report in 2021 (with greater confidence assigned to the transition ranges; see Figure SPM.4, panel a, IPCC 2022).

Although the Burning Embers has been a prominent visual in climate science and policy debates for a long time, it is less visible beyond science-policy spaces. This chapter now turns to two scientific visuals which have 'gone viral' to feature much more prominently in public life.

Climate Spirals and Stripes

In 2016, a flurry of emails between three climate researchers with an interest in communication, two of whom had never met, documented the development of a new scientific figure designed to represent changes in global average temperature. Although global mean temperature is often visualised in scientific figures (indeed, it is a concept which haunts international climate policy negotiations, despite being a poor indicator of planetary health; Victor and Kennel 2014), the trio were motivated to visualise the evolution of global temperatures in a more engaging and dynamic way (Hawkins, Fæhn and Fuglestvedt 2019). Their collaboration resulted in a scientific image posted to the social media platform X (then known as Twitter) (Hawkins 2016).

Ed Hawkins, Taran Fæhn and Jan Fuglestvedt had collaborated to produce the now-iconic Climate Spiral. A moving image representing changes in global average temperatures, the Climate Spiral caught people's attention and quickly went viral. In this animation, a clock-like circle (but with months instead of hours) charts the regular monthly progress of global mean temperatures each year from pre-industrial times to the present day, via a year counter. Circular bands mark the climate policy boundaries of 1.5°C and 2.0°C. The climate policy bands are coloured red, to signal danger thresholds.

The gif plots the global temperature for each month in a given year. Because of the inexorable rise in global mean temperatures, the animation gradually spins outwards to the edge of the circle, appearing to increase in speed as it does so. The Spiral starts off purple, before progressing to blue, then green, ending in a deep intense yellow. These colours were chosen to visually signal that 'dangerous' levels had not yet been reached (Hawkins, Fæhn and Fuglestvedt 2019); a nod to the UNFCCC 2015 Paris Agreement, the legally binding international treaty on climate change intended to limit global mean temperature increase to 1.5°C above pre-industrial levels (UNFCCC 2023).

The original tweet has been viewed several million times, but a much wider audience for the visual was ensured when a version was shown on a large screen at the 2016 Rio Olympics opening ceremony; reaching an estimated two billion people and leading its creators to claim it was 'probably the most watched broadcast about the climate ever' (Hawkins 2016). The virality of this moving scientific image and its use in popular culture are indicators of its appeal; which are also speculated on by the authors in Hawkins, Fæhn and Fuglestvedt (2019). As yet, there is no empirical work with audiences which seeks to understand how people understand and engage with this most viewed of climate visuals. However, the success of the Climate Spiral was a precursor to Hawkins designing another visual that has also had huge popular uptake, the Climate Stripes.

In 2015, Marine Scientist Joan Sheldon started to crochet what she named her 'globally warm' scarf (Sheldon 2017), presenting the finished scarf at the Coastal and Estuarine Research Federation meeting in the USA. Her scarf is mostly purple, with some blues, and deeper reds towards the late twentieth century representing the warming climate (Schwab 2019). A couple of years later on the other side of the Atlantic, and independently of Sheldon's globally warm scarf, another keen scientist-crafter had a similar idea: to create a striped blanket based on a global temperature dataset. Climate Scientist Ellie Highwood crocheted a blanket for her climate scientist colleagues Jennifer Catto and Duncan Ackerley, as a gift for their new baby. Ellie's global mean temperature blanket was based on 100 years of temperature anomaly data. It featured a much wider range of colour than the globally warm scarf: ranging from dark to light blues, greens, to pinks, purples and yellows. Ellie posted a photo of her blanket to social media platform X (then known as Twitter), where she was astonished to find it receive more than a thousand likes and retweets (Highwood 2017, 2017b). So-called 'temperature blankets' are popular among crochet and knitting communities worldwide, where they can be an important tool for climate engagement through craftivism (Greer 2014, Moreshead and Salter 2023). Jen describes the gift (see also Figure 5.2): 'This is a much-loved gift FROM our colleague and friend Ellie; and something that is still in daily use years later! It's amazing to think of its story, and how it helped inspire Ed to

Figure 5.2: The crocheted Climate Stripes blanket, inspiration for the Warming Stripes. It was a gift to Rose, daughter of climate scientists Jennifer Catto and Duncan Ackerley.

Source: Jen Catto

create the Climate Stripes' (Jen Catto, Climate Scientist, Exeter University, personal communication, 8 December 2023).

Some months later, Ed Hawkins, creator of the Climate Spiral, was invited to speak at a Welsh literary event, the Hay Festival. He mused over how he could visualise, in a simple way, global mean temperature rise. He remembered chatting to his University of Reading Meteorology Department colleague, Ellie, about the blanket she was crocheting for Jen and Duncan's new baby, and in his words: 'the Warming Stripes were born!' (Ed Hawkins, pers comm, 26 January 2024; note the visual is also commonly called the 'Climate Stripes'). He decided to forgo standard science visualisation rules – for example, the figure lacks a title or legend – and instead relied on colour alone to carry the visual's message (Royal Society 2023). He used two ColorBrewer single-hue palettes (Dixon 2023): shades of red to indicate warmer than average temperatures, and blues to indicate cooler than average temperatures. Although there were far fewer colours in the Warming Stripes than Ellie's blanket, there were many more shades than Joan's scarf. As Ed describes, the graphic immediately started to grab people's attention:

> The version of the graphics shown at the Hay Festival was using temperature data for the town of Hay itself, with the idea of connecting climate change to people's actual experiences - local rather than global. As soon as I showed the graphic on screen, I could see people in the audience sit up and take notice. (Ed Hawkins, Climate Scientist, University of Reading, personal communication, 26 January 2024)

The Climate Stripes have since been described as a 'stunningly simple' way to visualise global mean temperatures (Royal Society 2023). Ed had created a figure which represented global mean temperatures in a way which was aesthetically pleasing, and yet still readable as a depiction of scientific data.

Buoyed by the interest in the Stripes, by 2019, Ed had created a website which gave a useable interface from which users could plot Warming Stripes for many countries and regions around the world (https://showyourstripes.info/). Warming Stripes were made downloadable for several hundred countries, regions, cities and places using data from organisations including Berkeley Earth, NOAA, the UK's Met Office and others; but consistency remained in terms of the same red-blue colour scale to visualise mean temperature changes. The Warming Stripes website was immediately popular, with over a million downloads in the first week alone (Reading University 2022). The Warming Stripes were released on a Creative Commons licence, which not only allowed reuse, but also remixing and transformation (Dixon 2023). This has resulted in a proliferation of their use and of Stripes creativity.

Unlike most scientific figures, the Warming Stripes have made their way into the everyday spaces of cultural politics, as well as formal politics (for example, Figure 5.3). The Stripes have been featured on catwalks to football kits, shower screens to nail art, rock band stage sets to public transport, garden fences to oil paintings (UEA 2023). In formal politics, they featured on the logo of the US House Select Committee on the Climate Crisis throughout its three year tenure (Wikipedia 2023), and have become a common visual, widely used at the COP climate summits and beyond. The Pope was even presented with a Stripes stole on World Environment Day in 2023, at an event designed to raise global awareness of climate change and the moral imperative to act (Giannoli 2023).

Figure 5.3: Reading Football Club display the Climate Stripes on their team shirt, during the 2022-23 season

Source: Reading University/Reading Football Club

On the Northern Hemisphere summer solstice each year (21 June), users across a range of social media platforms are encouraged to showcase their own inventive ways to #ShowYourStripes. This initiative began in 2018 when Florida weather forecaster Jeff Beradelli started #MetsUnite on the summer solstice, encouraging broadcast meteorologists to raise the profile of climate change in everyday spaces by wearing clothing featuring the Warming Stripes when presenting TV weather broadcasts, and through using Warming Stripes visuals in background graphics (MetsUnite 2023). #ShowYourStripes also creates a news hook for media organisations to feature climate change. Coverage has ranged from a climate awareness feature on Ugandan public broadcaster UBC (UBC 2023), to streaming service Netflix using the Stripes both to showcase shows and films with sustainability narratives and as a vehicle to advocate for a more sustainable media industry (Netflix 2023). They featured on the cover of Greta Thunberg's Climate Book (Thunberg, 2022) and the cover of *The Economist's* Climate Issue (The Economist 2019).

The simple message and appealing aesthetic of the Warming Stripes have also been remixed to start conversations on other issues, including the biodiversity Stripes illustrating rapid decline in biodiversity (Richardson 2023), to a remix of diversity in science via LGBTQ+ Pride colours (University of York 2022). They have arguably become 'one of the most

Figure 5.4: The 'Climate Generations' figure. Detail from IPCC AR6 Synthesis Report

The extent to which current and future generations will experience a hotter and different world depends on choices now and in the near-term

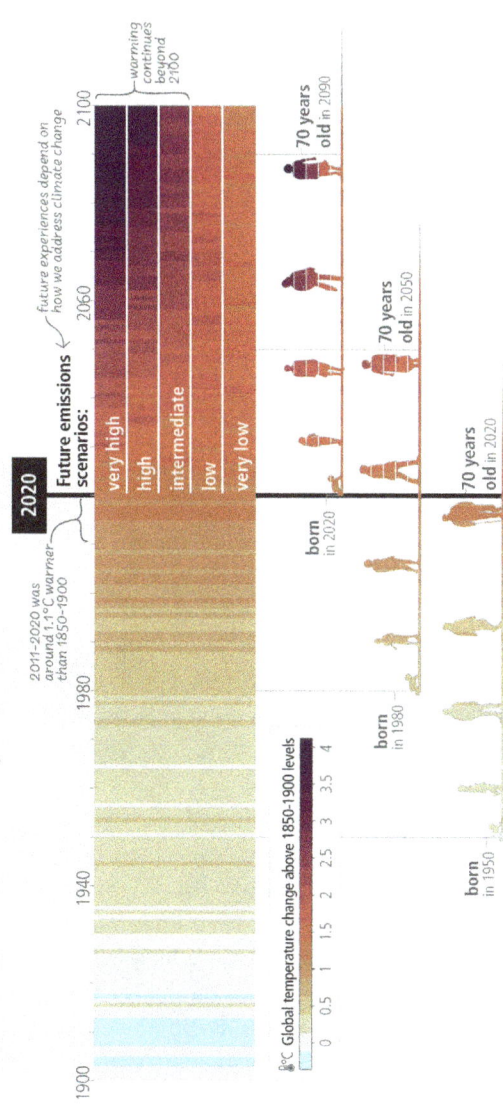

Notes: Original caption: 'Observed (1900–2020) and projected (2021–2100) changes in global surface temperature (relative to 1850–1900), which are linked to changes in climate conditions and impacts, illustrate how the climate has already changed and will change along the lifespan of three representative generations (born in 1950, 1980 and 2020). Future projections (2021–2100) of changes in global surface temperature are shown for very low (SSP1-1.9), low (SSP1-2.6), intermediate (SSP2-4.5), high (SSP3-7.0) and very high (SSP5-8.5) GHG emissions scenarios. Changes in annual global surface temperatures are presented as 'climate stripes', with future projections showing the human-caused long-term trends and continuing modulation by natural variability (represented here using observed levels of past natural variability). Colours on the generational icons correspond to the global surface temperature Stripes for each year, with segments on future icons differentiating possible future experiences.'

Source: IPCC 2023

iconic graphics of modern times' (Royal Society 2023). However, despite their widespread use, there is yet to be any empirical assessment of how people engage with the Climate Stripes, as a form of climate science visual. This assessment is important, as the Climate Stripes may not be interpreted as assumed. For example, one scientist was surprised to discover that the Warming Stripes blue/red colour scheme were not viewed as a record of temperature change; but as a party political message:

> I am a volunteer on my town's Environmental Committee, and suggested putting the iconic Climate Stripes image on our webpage. To my astonishment, one of the committee members recommended avoiding red/blue distinctions as the image colours have a political connotation: blue for the Democratic Party, and red the Republicans. I remain totally amazed that anyone viewing the Stripes would consider it to be a political swipe at Republicans! (Andrew Goodwillie, Geophysicist, Columbia University, personal communication, 14 July 2023)

This is not an isolated finding. McMahon, Stauffacher and Knutti (2016) designed a study to investigate how people – both climate experts and non-experts – responded to climate scientific visuals, including four IPCC visuals. Post-test interviews indicated that climate expertise was important when decoding colour scales. A non-expert participant, for example, interpreted the blue and pink colour scale as relating to birth rate of males and females, rather than temperature anomalies. The red-blue colour scale for temperature changes may not be as self-evident as is assumed, and absolutely underscores the importance of testing how visuals are read and interpreted with target audiences.

The Warming Stripes were incorporated into the scientific-policy boundary at the most visible level; through their inclusion from the IPCC's Working Group I report in 2021 into the IPCC's Synthesis Report, the summary document for the entire Sixth Assessment cycle (Figure 5.4; IPCC 2023). Here, the visual design of the Warming Stripes was modified to add human figures, to change the timeline and to change the colour scheme.

The 'Climate Generations' visual builds on the Warming Stripes by portraying not only historical global mean temperature changes, but by splitting the second part of the graphic into five different bars; where each represents future projections of global mean temperatures under very low to very high emissions scenarios. This drew on the 'warning Stripes' work of German engineer Alexander Radtke (Thompson 2019); as well as US climate scientist Alex Ruane, where he aimed to show how taking action on climate change impacted global mean temperature projections (Hawkins 2023). The 'Climate Generations' graphic adds human figures underneath, to visually link the changes in global mean temperature to people's lifespans, depending on whether they were born in 1950, 1980 or 2020. This addition was inspired by designer Arlene Birt, who had experimented with adding information about human lifespans to IPCC graphics during an artist residency prior to her role designing the figure as graphics officer of the IPCC Synthesis Report (Background Stories 2024). Her intention was to try and relate seemingly abstract data about climate futures to everyday people and their families: she comments how she experienced an emotional reaction when developing the concept, as the icons align with her mother, daughter and herself:

> When I mapped the generations of women in my own family, I came to tears realizing how my daughter and I will experience a very different world — much different than my grandmother and great-grandmother. This figure helps us realize how future generations of our families will live in very different climate-worlds than current and past generations. (Arlene Birt, Founder and Creative Director, Background Stories, personal communication, 8 April 2024)

Finally, the graphic departs from the simple two-colour blue-red Warming Stripes colour scheme, to one which also encompasses a controversial purple (discussed further later). The Climate Generations figure as a whole is an example of how the IPCC has moved to a process of co-design, incorporating insights from cognitive scientists, information designers as well as subject content experts to make the IPCC report visuals more user-friendly (see Morelli et al 2021).

The Climate Generations figure has sparked considerable attention since the IPCC's 2023 Synthesis Report was released; garnering positive reviews of its ability to visually portray the impact of climate warming on present and future generations; including in the UK's *Financial Times* (Campbell and Bernard 2023), US *Washington Post* (Patel 2023), as well as at *Le Monde, Corriere Della Sera, NBC* and *Axios*. It even evolved into memes posted on social media (for example, see ClimateShitpost 2023):

> This figure is one of the few in the history of IPCC to be widely shared by global media, unaltered and without the need for clarification for non-specialist audiences. My favorite examples include local meteorologists using this figure to share the futures of their own children. This demonstrates the intended effect of making the typically abstract concept of future climate change more psychologically close (more 'real') to non-expert audiences. (Arlene Birt, Founder & Creative Director, Background Stories, personal communication, 8 April 2024)

There have also been suggestions for edits, such as from WWF scientist Stephanie Roe to include icons of non-humans, as well as icons of people (Roe 2023); and criticisms, including from NASA scientist Chris Colose, who suggested the figure was too complicated (Colose 2023). A further development has seen NASA develop an interactive version of the figure, which allows users to add people's lifespans of relevance to them, such as their own relatives (NASA 2024). As with the Climate Stripes, there has not yet been any empirical work to understand how audiences engage with the Climate Generations figure.

Weather maps

As alluded to before, a recurring theme in the public discussion of climate science visuals is the role of colour; or, to be more precise, the affordances (meaning-making) that colour can carry – which communicators select according to their needs, interest and context (Kress and van Leeuwen 2020). These affordances are partly via the material characteristics that define colours, which run on a series of scales: of hue (what is commonly referred to as colour: red, blue, green, yellow and so on); value (the lightness or darkness of a colour, from maximally light (white) to maximally dark (black)); and of saturation (from 'pure' colour to 'pale' or 'pastel', and ultimately draining to just values) (Kress and van Leeuwen 2020, Rose 2023). Second, colour affordances come via association: 'where have I seen this before?' – and therefore an interpretation of what the use of a colour means, that is, its symbolic value within a particular sociocultural context (Kress and van Leeuwen 2020).

Throughout the scientific climate visual examples discussed, colour plays a key role in meaning-making. In the Climate Spiral, the Spiral's hue starts off blue, before turning green and yellow, but not red; to indicate that a 'dangerous' level of climate change was not yet reached (Hawkins, Fæhn and Fuglestvedt 2019; that is, global mean temperature increase had not yet reached the 1.5°C guardrail of the Paris Agreement; UNFCCC 2023). In the Warming Stripes, scientist Andrew Goodwillie reads the Climate Stripes red and blue hues as relating to temperature, and is astonished that they might alternatively be read as a US party political sign. In the Climate Generations figure, purple hues are featured in the scenario bars; and the use of purple in visually representing climate science data has proven contentious before (Dixon 2023). Climate controversy over colour has been bought to the fore particularly through the last type of image to be explored in this chapter: the weather map.

Climate scientist Julie Arblaster relates in an interview (Tamman 2021) how purple started to feature on the Australian Bureau of Meteorology (the Bureau) weather maps from 2013. She narrates everyday discussions she had with her meteorologist partner, Jim Fraser, as they were both then working at the Bureau (see also AMOS 2023). Historically, the highest recoded temperature in Australia was set at 50.7°C in 1960; and hence the maximum extent of their weather map scale was the 48–50°C range. The top part of the scale, through 40°C, was a graduated scale ending at deep reds. But, during a heatwave in what became known as the 'angry summer', the Bureau models predicted higher temperatures than this top point of their existing scale. So, above 50°C, the chart would display as without a colour, as white. Julie described how her partner Jim and colleagues at the Bureau decided that the colour white to indicate record-breaking temperatures did not work graphically, and so he instead quickly picked purple, and the forecast was put out (Figure 5.5).

The news that Australia needed a new colour for its weather maps because of unprecedented heat quickly spread around the world's news media. Climate contrarian blogs made much of the colour purple; suggesting that it was a conspiracy rather than an everyday workplace decision made by the meteorologists, albeit in the context of an extraordinary weather event. Julie explains:

> To my knowledge, the choices being made by the operational forecasters are very much removed from any conscious climate communication, i.e. the forecasters aren't actively trying to make the link with climate change, but the data is somewhat doing it for them. The decision to add another colour to the chart was purely a solution to an operational issue. It was climate scientists that actually made the link and promoted it on social media. My choice – of binding my PhD thesis I submitted that year in an unconventional purple cover – was in some ways a tongue in cheek recognition of this. For all my climate change research, Jim's simple act of adding a colour to a weather map did much more to communicate the urgency of climate action; it really struck home in that Angry Summer of 2013. (Julie Arblaster, Climate Scientist, Monash University, personal communication, 12 January 2024)

Purple had arrived in temperature charts, and the colour began to act not only as a signifier for extreme heat, but also as a signal for climate controversy. The significance of these events has subsequently been recognised through an AMOS Award to Jim Fraser. As the AMOS website states, he 'famously extended the colour palette of Bureau ACCESS charts to include two shades of "deep purple" for temperatures above 50°C' (AMOS 2023). It has also been recognised in the National Museum of Australia archives, as a defining moment in Australian history (National Museum of Australia 2023).

Environmental Humanities scholar Deborah Dixon describes the use of purple by the Bureau as a visual disruption and a 'breach' (Dixon 2023):

Figure 5.5: Purple began to feature at the top end of the Australian Bureau of Meteorology's weather map temperature scale from 4 January 2013

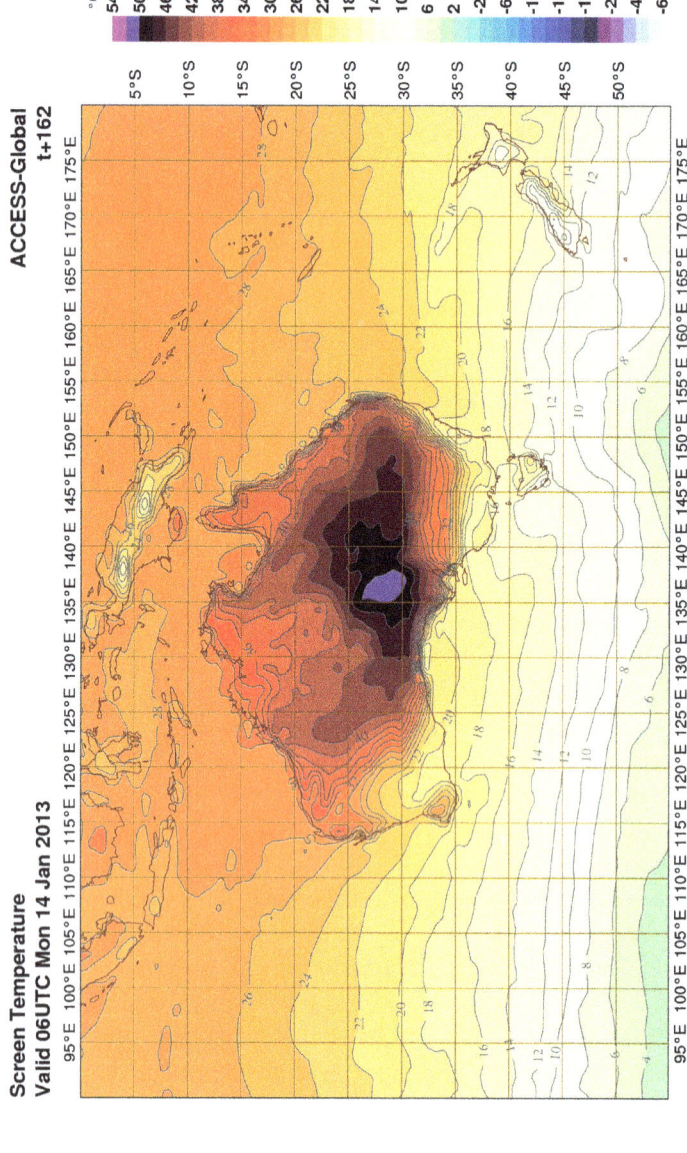

Source: © Commonwealth of Australia 2013, Australian Bureau of Meteorology

both of an everyday object; the banal and everyday standard measurement scale of the weather forecast; but also, by the link between the extreme heat projected by the weather map and the underlying science of extreme event attribution linking the 'angry summer' to climate change (Lewis and Karoly 2013); that is, it signalled a breach of what is imaginable for global climate.

Controversy, indicating a breach of weather forecast expectations, has continued apace since. The UK's Meteorological Office (Met Office) has modified the way it visually presents forecasts over time to improve forecast communication. For example, because of technological advancements, the Met Office changed from using symbols on weather maps, such as a temperature reading in the centre of a square icon, to using graduated scales of colour to represent expected temperature over a whole land region. And, to make their forecasts easier to read for people who are colour blind and others, the Met Office now uses fewer hues in their colour scale, and instead makes greater use of values from light to dark (Met Office 2023). The Met Office temperature scale is static; that is, the same colours are used for the -55°C to +55°C scale in every region and season (Met Office 2023).

In the UK, temperatures generally sit somewhere between 5°C and 25°C through the year, which results in hue and value transitions from light yellow and into orange and light red (Met Office 2023). In summer 2022, the UK breached 40°C for the first time in meteorological history. This resulted in much darker colours being featured on weather forecasts. These unexpected colours – this breach of weather forecasting expectations – resulted in substantial controversy, including the viral spread of the meme in Figure 5.6 over summer 2022. The meme, 'It's called summer', uses an old-style Met Office map (using symbols) on the left, and another Met Office map on the right, to spread misinformation; by implying that the Met Office has changed its forecasts to generate 'fear' rather than because of technological and communication advances. As Met Office meteorologist Aidan McGivern explains:

> The right hand map was taken from a tweet that the Met Office put out in 2016. This was when the Met Office first started using new graphics software called Visual Cortex and was experimenting with different styles and approaches. Soon after, we settled on a much more consistent colour scale, which we used for a few years before more recently adapting it for accessibility reasons. (Aidan McGivern, Meteorologist, Met Office, personal communication, 11 December 2023)

As more extreme temperatures occur more often, these colours, and the controversy they engender, have become a persistent and recurring trope in the right-leaning UK media. A prolonged heatwave in southern Europe in summer 2023 saw these colours being used consistently in weather

Figure 5.6: 'It's called summer' meme

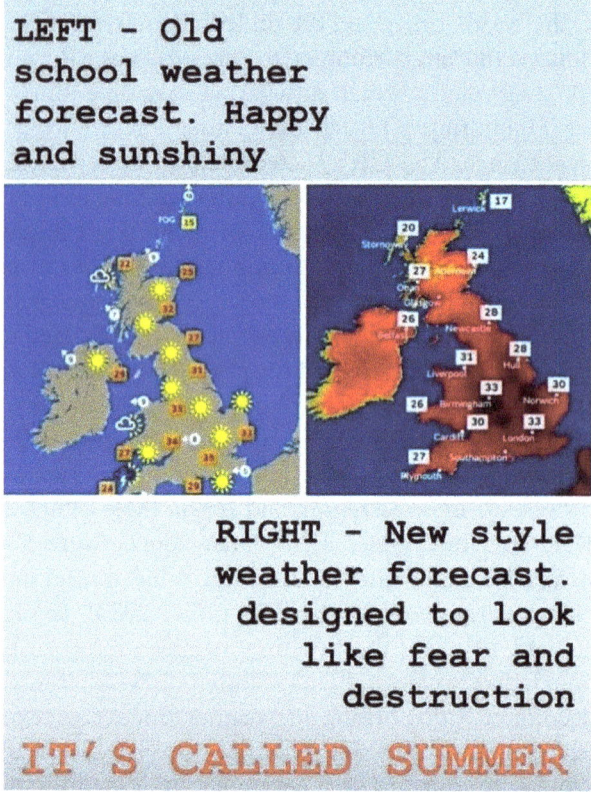

Notes: The meme went viral in July 2022. The right-hand side weather map is a map from the UK Met Office that was taken out of context to spread climate misinformation. Met Office meteorologist Aidan McGivern responded to the meme on Twitter.

Source: Meme featured in McGivern 2022 explanatory tweet thread

forecasts. A segment from UK TV channel GB News saw presenter Neil Oliver making a series of false claims of the BBC and others 'pushing climate terror' through colours on weather maps, in a clip viewed more than two million times (Silva 2023): 'when did we get to the point where every weather map through the summer is in shades of red and black ... creates this illusion ... turning the planet into what looks like an overcooked pizza ... blistering and black and bubbling as though the place is melting ... and its blatant and patent nonsense' (GB News 2023). The clip then goes on to make a number of false claims about the science of climate change (claims which variously fit within the taxonomy of contrarian misinformation, see Coan et al 2021).

A similar sentiment was broadcast on opinion-orientated TV channel TalkTV, 'Julia blasts the "hysterical" reporting of the European heatwave', as a satirical and sarcastic conversation between anchor Julia Hartley-Brewer (JHB) and guest Rupert Dahwall (RD) (Talk TV 2023):

JHB: What have you made of, well it is – hysterical – reporting of this heatwave? …
RD: Well, what I'm really impressed by is the weather maps, and the new, new, uhrrm, spectrum of colours they've bought to it, new shades of purple verging on black and …
JHB: and not just red, I mean it's not just from red, green, yellow, amber, red … it's purple and black 'that means it's very bad'!!
RD: Yes, as you say, it means … it means 'we're all going to die!' from the heat.

This narrative was also present in the right-leaning print media. Columnist Allison Pearson wrote in the UK's Telegraph newspaper:

> What is the betting that the sudden change in TV weather maps, from pastoral greens and shy yellows to diabolical reds, even bruised purples and black, was suggested by the Nudge Unit? [the UK Government's Behavioural Insights Team] … They have over-reached this time, I think, with their lurid weather maps. (Pearson 2023)

Note the use of emotional terms Pearson satirically employs to describe standard weather forecast use of hues: the simplicity and historical charm connoted by the rural 'English pastoral' (see Rebanks 2020) and 'shy' yellows and greens; compared to dangerous and seemingly tricksy 'diabolical' and 'bruised' reds, purples and blacks.

Misinformation about the forecast became so persistent that both the Met Office and the BBC put out explainers articles: 'No need to see red over Met Office colour scale' (Met Office 2023) and 'What do colours on the BBC Weather maps mean?' (BBC 2023). Further to this, the BBC followed up their original article with a special report factchecking various claims of misinformation, via its Climate Disinformation Reporter (Silva 2023):

> As part of my job, I have witnessed how imagery can be used to great effect by those spreading misinformation about climate change. In 2022, I reported how some social media users dismissed record-breaking temperatures across the UK, by comparing them to those registered in the summer of 1976. Those users wrongly claimed that the hot weather was being exaggerated by weather forecasters – and, to back their point, some of them shared old photos of people eating ice creams, enjoying the beach, or bathing in public fountains. Those images were meant to appeal to a sense of nostalgia. But those users

ignored the fact that the peak temperature in 1976 was 35.9C, as opposed to 40.3C recorded on 19 July 2022. By posting these photos alongside memes (often critical of the supposedly 'dramatic' colours used in modern weather maps), they tried to undermine suggestions that the extreme weather was unusual and made more likely by climate change. (Marco Silva, senior journalist, BBC News and the BBC's climate disinformation reporter, personal communication, 8 May 2024)

The association between colour and temperature (red as hot, blue as cold), then, is contingent and not universal (Dixon 2023). Interpretation of colour has differed substantially over time and across cultures. For example, as it was the first colour to be materially developed for painting and dying, red was associated in western antiquity with wealth and power; before moving to be associated with the fire of Hell in the Christian religion during the medieval period; then as a symbol of excess and immorality during the Protestant Reformation, before being linked to radical left-wing politics after the French Revolution (Pastoureau 2016). While today, red may well be indelibly linked to danger in western cultures, it is an auspicious colour in China, where it is the colour of life, good luck, joy and happiness (Huang 2011), and used widely in celebratory occasions such as weddings. Just as any other part of an image, colours are 'read' (Hall 1997): the meaning of colour is not static nor self-evident, but works to carry social meaning (Aiello and van Leeuwen 2023).

Where to next, for climate science visuals?

This chapter has explored some of the most iconic scientific images of climate change. The examples in this chapter challenge any assumption that scientific images are somehow different to other types of climate visuals, in terms of their objectivity. Just as any other image, climate science visuals are a representation of reality and work as normative statements about how the world works (Urry 1992, Hall 1997). Indeed, scientific images work reflexively through the act of representing possible futures – making some types of future more or less likely - implicitly becoming political images (Schneider and Nocke 2014) as they develop a life 'beyond the lab'.

As Harold et al (2016) discuss, visually representing climate data can be tricky, due to a wide diversity of audience needs and interpretation skills (Lorenz et al 2015), the multi-dimensionality of data and challenges posed by the software and tools employed to create graphics (Nocke et al 2008). Yet the widespread use of climate science visuals such the Climate Stripes and Spiral, and their creative reuse and remixing such as through the Climate Generations figure, clearly attest to their popularity and aesthetic appeal. The UK's Royal Meteorological Society writes that 'the Warming Stripes are already having a massive impact' (Plumb 2023). In the same article, the Royal Met Soc reiterates Ed Hawkins' aim that the Warming Stripes are shared by meteorologists and scientists to 'start conversations about climate change'. The article also suggests that the Stripes 'promote awareness of the climate crisis, and stress the urgent need to take action now' (Plumb 2023: np).

But what does it mean by the Warming Stripes having a 'massive impact'? What of a deeper understanding of how people engage with the Warming Stripes, and their derivatives such as the Climate Generations figure? Do they start conversations, as is assumed by the earlier quote? Who has these conversations? One of the first rules of effective engagement is to 'know thy audience'; and thus recognising that people have different psychological, cultural and political reasons for (not) taking climate action (Yale Program on Climate Change Communication 2023). At the Yale Program, a series of audience segmentation studies has sought to understand US engagement with climate change. Their 'Six Americas' work, for example, suggests the Climate Stripes might reach the 'Alarmed' or 'Concerned' segments of US society; but what of the 'Cautious', 'Disengaged', 'Doubtful' or 'Dismissive'?

In terms of these conversations that the visuals may spark: what do these discussions look like? What, if anything, is the impact of seeing the Warming Stripes in terms of a broader set of questions around climate change engagement: on values, attitudes or behaviours? A question unanswered at present is whether the intended scientific message is interpreted as scientists assume.

In a novel study of user engagement with climate science visuals, Lorenz et al (2015) tested the effectiveness of scientific visuals showing climate projections with both UK and German government decision-makers tasked with adapting to climate change (that is, a key target audience for this sort of visual information). They found participants' comprehension of the information conveyed differed substantially both between participants, and across different graphical formats. They also did not find a consistent association between participants' comprehension of the figures (how well they actually understood the information presented, as assessed by independent testing) and their perceived comprehension (how well they thought they understood the visual). In a similar study, Fischer et al (2020) tested how political decision-makers and junior diplomats (that is, IPCC key audiences) interpreted two figures from the IPCC Fifth Assessment Report, finding that, when the graph was presented non-intuitively, the majority of participants misinterpreted it and drew the opposite conclusion from what was meant to be conveyed. Not only was it misinterpreted, but these participants also had high confidence that they had interpreted the figure correctly (when they had not). In another study, this time, an experimental approach using eye-tracking technology, Libarkin, Thomas and Ruetenik (2013) found that while scientists generally understood where to look for key information in climate science visuals, non-experts tended to ignore information like the figure's legend and scale, and focus instead on artistic yet unimportant visual cues, in random order. And as a final example, in a US study of how people interacted with maps showing projected flooding in local communities, Mildenberger and colleagues found that people viewing the maps – even where the maps showed flooding of their own homes – acted to reduce concern about future sea-level rise (Mildenberger et al 2024). These studies suggest that assessing how different audience groups engage with climate science visuals is complex and challenging; rather than intuitive or obvious.

Science literacy is not universal, and indeed, many people struggle to understand or interpret scientific figures, including simple climate change graphs (Shah and Hoeffner 2002, van der Linden et al 2015, Fischer et al 2018). Climate science figures can also make climate change seem less relevant to people's everyday lives: in an experimental study, Duan, Takahashi and Zwickle (2019) showed how abstract climate change images, such as scientific graphs (in contrast to concrete images, such as photographs of climate impacts) were more likely to make viewers perceive climate change as a spatially and temporally distant issue. The astonishment in the quote from scientist Andrew Goodwillie, about the unanticipated reaction to using the Climate Stripes, demonstrates how – though valuable as a historical record of the development of an iconic graphic – articles written by scientists (for example, Hawkins, Fæhn and Fuglestvedt 2019 on the use of the Climate Spiral) can only hypothesise on why and how people are interpreting and engaging with climate science visuals; and thus highlights the need for social science insights into climate visualisations, and co-design approaches which bring together different types of expertise such as graphic design, data visualisation and cognitive science (Morelli et al 2021).

Given the increased presence of climate science visuals in public life, and the growing controversy over seemingly innocuous (to scientists) visual cues such as the presence of purple in a climate science visual's colour scheme, these insights would appear to be increasingly important. Empirical social science research seeking to understand public engagement with these iconic climate science visuals would therefore appear to be an essential, yet largely missing, part of climate science communication to date.

6

People: Politicians and Protesters

In many climate visuals, and certainly in news coverage of climate change, images depicting people are very common (O'Neill 2013). Images of people are also prevalent in climate content online – from Twitter posts, to viral videos on TikTok, to popular climate change memes (León, Negredo and Erviti 2022, Hautea et al 2021, Ross and Rivers 2019). Conversely, scholars have found that in some forms of online communication, in Google Image search results, images depicting people are almost entirely absent (Pearce and De Gaetano 2021). And who are the people depicted, when climate images do feature people? From climate news to Facebook adverts, politicians appear to dominate, but everyday people are largely missing (León and Erviti 2013, O'Neill 2013, Weaver et al 2022). Overall, it is clear that the historical representation of climate change as an environmental issue separate and discrete from humans and culture remains problematic (Doyle 2011).

This chapter begins by focussing on the types of people often portrayed in visual representations of climate change, explaining how political figures dominate and how other types of people who we might expect to see in climate coverage (such as celebrities, scientists, business leaders or people impacted by extreme weather) feature far less often. A discussion of political leaders featured in climate visual news coverage demonstrates both how the journalistic norm of personalisation, and also the scheduling of major political events, come to shape visual coverage of climate change. This provides a segue to a closely aligned visual trope, that of climate protest; where the discussion demonstrates how, much like the polar bear visuals in Chapter 3, representations of climate protest visuals have changed over time – from a display of social deviance, to a visual portrayal of protest as intergenerational justice (Hayes and O'Neill 2021). Finally, the chapter looks to an important part of Internet culture, the meme; where images of identifiable people are morphed and riffed on to provide humorous and ironic social commentary, as well as to signal alignment with different value positions on climate change (Ross and Rivers 2019).

Political figures

Identifiable people are a common type of visual used to represent climate change. As much as half of the coverage in a 2010 study of US, Australian and UK newspapers featured images of identifiable people (O'Neill 2013); a finding consistent with a study of Canadian newspaper coverage (DiFrancesco and Young 2011) as well as in US news articles (Rebich-Hespanha et al 2015). A similar pattern exists on some social media platforms, with a more recent study of climate imagery on Twitter (now known as X) finding that a majority of posts contain pictures of people, with a third of all images containing identifiable people (León, Negredo and Erviti 2022). Perhaps this is not surprising, as it is the playing out of the journalistic norm of personalisation, or what Bennett (2011: 45) describes as the 'overwhelming tendency to downplay the big, social, economic, or political picture in favour of the human trials, tragedies and triumphs that sit on the surface of events'.

Perhaps what is more surprising, at first glance, is the type of people who feature in climate news visuals. Images of politicians and other people associated with politics feature particularly prominently in news coverage. For example, somewhere between a quarter and a third of all UK and US newspaper visual coverage of climate change features a political figure (O'Neill 2013, Rebich-Hespanha and Rice 2016). In contrast, images of other prominent figures in the public life of climate change, such as scientists, celebrities or business leaders are much less common, although this does vary by the news organisation: more coverage of scientists in the UK's left-leaning broadsheet *The Guardian*, and more coverage of celebrities in the UK tabloids *The Sun* and *The Express* (O'Neill 2013, Hayes and O'Neill 2022). Newspaper ownership also appears to be a factor in influencing the amount of visual coverage of political figures, with News Corporation-owned titles significantly more likely to picture political figures than those under other ownership (O'Neill 2013). Perhaps most surprising of all is the low coverage of everyday people impacted by climate change: León and Erviti's study of Spanish TV news coverage found just 5 per cent of TV news stories were accompanied by such visuals (León and Erviti 2013).

The large quantity of political figures in climate news visuals is likely explained at least in part by newsroom structures and routines. The politics desk ('desk' refers to department or staffing arrangements within a newsroom) is often much bigger than others, such as the climate and environment desk; and so much 'climate' news is actually reported by the politics desk rather than environment or climate specialists. This can influence, in turn, the type of visual coverage a story receives. For example, as a story is reported through the politics desk, it may be more likely to be viewed and reported through the lens of political conflict (Marcinkowski 2014); a situation often seen in climate coverage through the reporting of major political summits like the annual United Nations Framework Convention on Climate Change (UNFCCC) Conference of the Parties (COP). Photographs of politicians are therefore used to situate political events 'through the category of the subject', that is, politicians (Hall 1973: 229). Also, as much climate news reportage is driven by large-scale political events rather than weather or climatic characteristics (Schäfer, Ivanova and Schmidt 2014, Hase et al 2021), the regular occurrence of major international climate summits like the UNFCCC COPs are another factor driving the production of climate news, and onwards to a visual focus of political figures. So what does visual coverage of political figures in climate news look like?

In 2010, I was living and working in Australia, and observing first-hand the fractious state of Australian politics, and climate politics particularly, which followed through to visual media coverage of climate change. A leadership spill, caused partly by climate politics (though a delay to a proposed emissions trading scheme), saw Deputy Leader Julia Gillard sworn in Prime Minister and then contest that position shortly afterwards in a Federal Election. This resulted in a hung parliament with Gillard's Labour Party in minority government, and the Green Party holding the balance of power. The Greens had risen to prominence and picked up votes in the election partly due to dissatisfaction with the Labour party's lack of progress on the emissions trading scheme.

Figure 6.1 is a powerful image to illustrate this situation, a photograph originally published in the left-leaning Australian newspaper *The Age* (Grattan 2010). The dark background of the photo focusses attention on the eyes and faces of the two politicians pictured: Green Senator Bob Brown and the Labour Prime Minister Julia Gillard. In their book, 'Image Bite Politics', Grabe and Bucy (2009) devote a chapter to 'Facing the Electorate', where they use a range of evidence to argue that facial displays are influential elements of political imagery. Early research demonstrated how facial displays of leaders impact the emotions and attitudes of viewers (Masters et al 1986); with more recent work showing how factors such as a leader's gender will impact audiences interpretations of their facial expressions (Boussalis et al 2021).

In Figure 6.1, Brown's head is turned towards Gillard; compelling the viewer to follow his gaze. Brown's eyes stare at Gillard, his mouth is closed. The photographer's telephoto lens blurs the background, and emphasises the closeness of Brown sitting next to Gillard. Given the electorate's familiarity with the climate-focussed backstory, as well as the fractious nature of Australian politics and the dependent relationship between Brown and Gillard in a hung parliament, the viewer is left in no doubt that Brown is 'keeping a close eye' on Gillard's decisions regarding climate policy. More broadly, this image is a good example of how climate visual politics draws heavily on the attributes conveyed via facial expressions to both personalise climate news

Figure 6.1: Typical image featuring political figures

Notes: The photograph was originally printed in Australian newspaper *The Age*, 28 September 2010. It shows Labour PM Julia Gillard being closely watched by Greens Senator Bob Brown during a news conference for the new climate committee's discussion on carbon pricing.

Source: Andrew Meares, *The Sydney Morning Herald*. Reproduced with permission

(by focussing on the human trials, tragedies and triumphs of political figures negotiating on climate change) and to dramatise climate news (by simplifying complex narratives into clear and recognisable stories; Bennett 2011).

As discussed earlier, climate news reporting is partly driven by large-scale political events, rather than by, say, extreme weather events – as one might perhaps expect (Schäfer, Ivanova and Schmidt 2014, Hase et al 2021). A regular peak in climate news occurs around the annual UNFCCC COP, hosted by a different nation in November/December each year. Some COPs are more newsworthy than others, partly based on the status of the annual UNFCCC negotiations: for example, COP15 at Copenhagen (nicknamed 'hopenhagen') was particularly newsworthy, as an international legal framework for climate mitigation to extend beyond the 2012 end-point of the Kyoto Agreement needed to be negotiated (COP15 was additionally and unexpectedly newsworthy, because of the hacking and release of scientists' emails from a server at the University of East Anglia's Climatic Research Unit during the COP15 negotiating period, an event which became known as 'Climategate'). Conversely, in other years, the complexity of negotiations (such as in political wrangling over Loss and Damage) and for COPs more focussed on renewing, rather than increasing, ambition on climate targets (such as at COP27 in Sharm-El-Sheikh), a COP can be less newsworthy.

Whether more or less newsworthy, though, visual news coverage is similar in style and tone, both across nations and across different COPs. Commonly, there is a focus on photographs capturing discussion, speeches and individual portraits (Eide 2012, Grittmann 2014), memorably described by Wozniak, Wessler and Lück (2017: 1436) as 'the dreary routine of "talking heads"'. A study of UK news coverage indicates that a majority of visuals depict people, and of these images, many of them (38 per cent) portray a political figure (Hayes and O'Neill 2024). The highly staged nature of the event (and thus lack of opportunities for visual creativity) results in many visuals also cuing an institutional reference (such as the UNFCCC or COP26 logo), as well as a predominance of the official colour hue of the UNFCCC, a deep, saturated blue (see colour wheel in Figure 6.2). This is notably different from other types

Figure 6.2: Colour wheel for images from COP26 (Glasgow) news coverage in 8 UK news outlets, plotted by saturation (x-axis) and hue (y-axis)

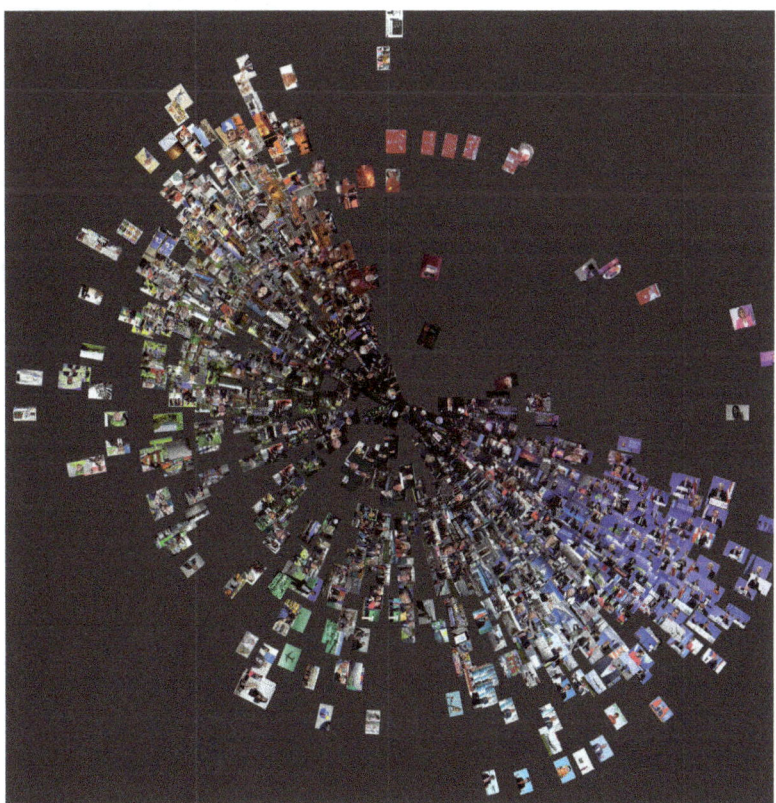

Source: Produced using ImagePlot. Previously published in Hayes and O'Neill, S. (2024), available CC BY 4.0

of climate visual coverage, such as climate impacts including wildfire and droughts, which feature many images in the orange-red-brown parts of the colour wheel.

The considerable similarity in COP news coverage across culturally very different nations (Wessler et al 2016) suggests a focussed news production event. Rules and limitations on media access, as well as on information supplies, constrict COP news visuals. My colleague, Geographer Sylvia Hayes, carried out a visual news ethnography at COP26 in Glasgow. She found a Getty Images photographer described the 'slightly overzealous' nature of UNFCCC security staff in negotiating access to the COP26 venue, for example (Hayes and O'Neill 2024). There is also an increased level of interaction and cooperation between normally more separate actors (including journalists, NGOs, political actors, photographers) for a time-limited, geographically co-located period (Lück, Wozniak and Wessler 2016, Wessler et al 2016, Hayes and O'Neill 2024) – or what Adolphsen and Lück (2012 in Lück, Wozniak and Wessler 2016) report journalists attending COP16 describing as a 'camp feeling'.

A typical example of the talking heads photographic style so common to COP visual coverage is provided in Figure 6.3. This image is from COP26 in Glasgow, UK, in 2021 and was taken by Kiara Worth, in her position as the UNFCCC's official photographer (in this role, Worth's photographs are all provided free-of-charge via a creative commons licence). It depicts an upper-body photograph of the COP26 host, UK Prime Minister

Figure 6.3: Typical image arising from a UNFCCC COP

Note: See how the podium, official UNFCC blue hue and flags are used as visual cues for the event surrounding the UK Prime Minister Boris Johnson at COP26, Glasgow, 2021.

Source: UN Climate Change, Kiara Worth, CC BY-NC-SA 2.0

Boris Johnson, speaking at the World Leaders Summit for the Opening Ceremony. Visual cues for a political leader are provided by the lectern and microphone, his business suit attire, and by the flags in the background (in this case, signifying the UNFCCC and the 2020–2021 country host, the UK). Visual cues for the UNFCCC COP process are provided by the two logos on the lectern (one for the UNFCCC process, the other for the COP26 specifically), as well as by the official blue hue of the UNFCCC dominating the background. Johnson's gaze beyond the photograph pulls the image's viewer into the perceived audience for his speech.

Hayes' interviews with photojournalists and editors at COP26 are instructive again, here. She found that, although these newsmakers recognised these sorts of talking heads images could be boring and repetitive, they were a necessary part of their role as documenters of the negotiations; that is, they performed the journalistic ideal of being a watchdog (Palmer, Toff and Nielsen 2020; see also Hayes and O'Neill 2024). Hayes' research also indicates that there is evidence of strict visual gatekeeping around who 'counts' as high-profile enough to be featured as a COP talking head. Perhaps partly because of the increase of non-environment or climate beat specialists being brought online to cover the COP, perhaps partly because of a perceived lack of public profile, newsmakers used visuals of heads of state preferentially to the COP26 President Alok Sharma (Hayes and O'Neill 2024).

Audience research can provide some insights into how people engage with these sorts of typical talking heads political figures climate imagery. In our Q-method study exploring UK, US and Australian public perceptions of mainstream climate news imagery (O'Neill et al 2013), we found images of people, and in particular, images of political figures, made people feel strongly both that climate change wasn't an important issue (that is, it provoked a low sense of saliency) and that they weren't able to do anything about it (that is, it engendered a low sense of self-efficacy). A comparable study carried out in Germany, Switzerland and Austria found similar results (Metag et al 2016). Taken together, these results suggest people (at least, in the western nations examined to date) do not find this style of talking heads political figures climate imagery – images which are widespread in climate news reporting – particularly engaging.

Protesters

Protest imagery appears to be relatively common in media coverage of climate change news, in both newspapers and on TV (León and Erviti, 2013, O'Neill 2013). However, in terms of the imagery that people hold 'in their minds' (or 'affective imagery' – the images that are brought to mind at the mention of climate change), it is much rarer. Psychologist Zoe Leviston and team found less than 1 per cent of people mentioned climate protest when asked to recall the first few images which came to mind when climate change was mentioned (Leviston, Price and Bishop 2014).

Climate protest imagery includes two broad types of visual representation. The first type is of performance-style installations and activities. Similarly to the talking heads political figures imagery described earlier, international political events can also drive this visual coverage of climate protests. The eye-catching installations and protests at the COPs, for example, have been successful at capturing widespread media attention (Doyle 2007, Eide 2012, O'Neill 2013, Wessler et al 2016, Wozniak, Wessler and Lück 2017). These installations and activities can be highly symbolic pieces of performance art which are produced in part to capture the visual attention of media audiences and raise the profile of the campaigners and their message (Wozniak, Wessler and Lück 2017, McGarry et al 2019).

A high-profile example of a piece of performance protest is provided in Figure 6.4. The photograph was taken in October 2009, as part of a staged event intended to capture global political attention ahead of the UNFCCC COP15 summit in Copenhagen and its post-Kyoto Agreement negotiations. In the eye-catching protest, the Maldivian government convened an underwater cabinet meeting of 15 people in the tropical waters of Girifushi, Maldives. The Maldivian President Mohammed Nasheed is seen signing a

Figure 6.4: Typical image featuring a performance-led protest. Maldivian President Mohammed Nasheed conducts a 30-minute cabinet meeting underwater in Girifushi, Maldives, to protest for urgent climate action ahead of COP15 in Copenhagen, 17 October 2009.

Source: Mohammed Seeneen/Associated Press/Alamy Stock Photo

document which stated: 'We must unite in a world war effort to halt further temperature rises. Climate change is happening and it threatens the rights and security of everyone on Earth'. The 'dreary routine' of photographing the talking heads of politicians is turned around into a fantastic spectacle; the ministers in their scuba gear, seated at tables, signing a document; but standing on a white sand seabed, surrounded by coral, darting fish and crystalline seawater. Here, rather than a dreary visual being attached post-hoc to a story, the image itself creates and leads the story. As the world's largest international multimedia news provider, Reuters stated at the time that it was an 'attention-grabbing event to bring the risks of climate change into relief before a landmark U.N. climate change meeting' (Omidi 2009). This demonstrates the expectation of the role this visual will perform in a globalised media ecosystem – the visual is intended to capture media attention and spread widely – bringing along with it the narrative of climate injustice and need for urgent action.

Another typical visual of performance protest are those enacted by the Red Rebel Brigade, linked to Extinction Rebellion (XR). XR place art and compelling visuals at the heart of their campaign ethos and strategy. Red Rebels figures are aesthetically arresting, with eye-catching red draped robes and distinctive face make up. Kat Brendel, one of XR's art coordinators, stated: 'our bold imagery is helping to change the conversations around climate change' (Brendel, quoted in Berhman 2019: np). The Red Rebel Brigade was the creation of Doug Francisco and Justine Squire from The Invisible Circus. Francisco describes how the Red Rebel Brigade was quite casually devised. For example, while they made an aesthetic choice to use the colour red, the costumes themselves came about via a callout to participants attending a workshop to create outfits for the forthcoming protest. Someone brought along eight bin bags of red velvet left over from a cabaret venue, which became the Red Rebel Brigade's red flowing robes. The meaning of the colour choice was not prescribed, but was partly defined after the public reaction: 'I didn't think about it loads ... We wrote our publicity from what people thought about it ... red was obvious [to audiences] for what the colour represented: references to 'stop', to blood, to emergency' Doug Francisco, artist, activist, and co-creator of the Red Rebels (personal communication, 22 March 2024).There are also other links to be made between the colour red and the danger of heat extremes, as connoted in a thermometer (see Schneider 2018; and also the discussion about colour in climate visuals in Chapter 6).

The Red Rebel figures have mask-like white-painted faces and red lips, which act as mime masks to help focus attention on the emotions (sadness, grief, love, empathy) that the figures portray, but are also somewhat otherworldly or ghostly (Benjamin 2019). They do not speak. The figures come together in groups and move in unanticipated ways: in contrast to the busy and sometimes even frenetic or chaotic pace of protest, they are slow moving, deliberate and mournful and so can transform the tension of a protest to visually construct very different emotions and mood (Francisco 2023; and as Doug said to me, 'the police just didn't know what to do with it' (personal communication, 22 March 2024). The Red Rebels have become an unanticipated symbol of XR and a compelling feature drawing the attention of the news media (see also Francisco 2023):

> The first time the costumes were used was at the XR protest in London [in 2019]. In the morning, reporters were zooming in on the stereotypical "eco protestors". But by the afternoon, that had all changed: the Red Rebels were everywhere, from Japan Today to the New York Times. It was so visual, photographers just loved it, papers loved it. People were taking so many photos, we couldn't move at points. The tone of writing seemed to have changed alongside it, from 'these annoying eco warrior protestors disrupting things', to the scale of the environmental crisis. (Doug Francisco, artist, activist, and co-creator of the Red Rebels, personal communication, 22 March 2024)

The Red Rebels were particularly visible via an XR 'Grief March' protest walk through central London. Images that were reprinted by many news organisations, including the BBC, show the Rebels pictured outside the Cabinet Office on Whitehall, London. The red and white Red Rebel figures stand in a long line, all of them with their eyes closed or looking down. Drawn on their outstretched palms is the XR symbol. Opposite the line of Red Rebels is a line of police officers, in fluorescent yellow coats and metropolitan police uniform. The police officers also appear calm, with their hands clasped in front of their bodies, standing still. The police officers' faces have a neutral expression. The effect of the two opposing groups is certainly unsettling, as the police 'unwillingly take part in the very choreography they purport to oppose' (Kennedy 2022: 86). This visual portrayal supports Extinction Rebellion's vision as a movement for non-violent civil disobedience; that is, the more sympathetic portrayal or protesters acts to reflect the framing of the protest movement itself (Hayes and O'Neill 2021).

While XR have been successful at capturing media attention through the creation of strong visuals, their aesthetic choices have not gone without critique (see White 2022, who writes of how XR have grappled with addressing climate justice through their visual strategy). Through the Red Rebels imagery specifically, their white faces and lack of visible skin, 'gender bending' red lipstick, exaggerated black-rimmed eyes and flowing robes and willingness to challenge police authority and actively seek arrest raise significant questions around race, gender and class – who has the privilege to actively seek arrest (Kennedy 2022)? And indeed to their wider stated mission of non-violence (White 2022)?

Mass protest

A second type of climate protest imagery is photographs of mass protests. These images may be shot from overhead (presumably via a tall building, before the now widespread drone journalism) or at street level, but often show urban streets filled with people. Figure 3.3 is indicative of this sort of protest imagery: it was a photograph I took as part of the Walk Against Warming protest in Melbourne, Australia in December 2009 and depicts both mass crowds, recognisable as a protest via the many waving placards held by protesters, as well as the foreground focus on the person in the polar bear costume, face obscured. These images act to provide a sort of visual 'proof' (see Doyle 2009) of a protest, by turning difficult to comprehend large numbers of people into a coherent visual narrative.

During the 2000s, mass protest imagery depicted protesters in a depersonalised way – their faces were obscured by masks or disguises, or were not visible because of their smallness given the distance of the shot. The contested nature of the visuals was clearly visually cued via the portrayal of two opposing groups of people: police (often single or pairs of men in formal poses: arms folded, still, in uniform and generally in riot gear, interacting with protesters in sometimes violent ways) and protesters (informally dressed, a mix of ages and genders where faces were depicted) (O'Neill 2013, Hayes and O'Neill 2021). These sorts of visuals are indicative of the 'protest paradigm', with protesters positioned as socially deviant (Chan and Lee 1984, Hayes and O'Neill 2021).

However, in the late 2010s, images of climate change mass protests appeared to shift. In 2018, 15-year-old Greta Thunberg began her campaign of 'Skolstrejk för klimatet' (School Strike for Climate), by protesting every Friday outside the Swedish Parliament. By 2019, and after invites to speak at global events such as the UNFCCC COP24 in Poland and the 2019 Climate Action Summit in New York, this had developed into a global movement involving millions of students. A study of one of the largest rallies, in Sydney, Australia, explored how the images on placards that young protesters created drew on culturally significant icons (such as a globe) to make connections to locally significant issues (though portraying it on fire with text about local bushfires). Common media representations (such as headshots of politicians) were used in satirical ways to leverage emotional responses of anger, amusement and empathy, and to evoke a collective demand for urgent action (see Figures 3 and 6 in Catanzaro and Collin 2023). In media visual coverage, photographs of the protests began to portray protesters as individualised people. Rather than mass crowd shots, there was an increased focus on photographing smaller numbers of people (for example, a group of people holding a banner; or a focus on an individual holding a sign) and in showing protesters faces more clearly (see, for example, Figure 6 in Hayes and O'Neill 2021).

The people depicted in climate protest visuals are also different: in contrast to the mixed-gender, older people depicted previously, climate protest visuals started to show predominantly young women and girls (Hayes and O'Neill 2021). Although it is likely more women and girls took part in climate protests towards the end of the 2010s, their depiction in resulting climate protests visuals is not a foregone conclusion: women are often under-represented in media coverage due to media organisations reinforcing patriarchal norms (Zoch and Turk 1998, Armstrong and Boyle 2011). It is also notable that the camera angle of these images often present the young protesters as empowered; note how the visual's eye-level composition is designed to place these young people in a position of power (Caple 2013, Kress and van Leeuwen 2020). Again, this is a deliberate photographic choice of a photographer to crouch down in front of the protesters and aim the camera upwards: there is likely a significant height difference between these young protesters and the likely male, adult photojournalists capturing these images (see World Press Photo 2018 for a discussion of the male-dominated workforce of photojournalism). This suggests news organisations are moving to visually portray climate protest as an issue of intergenerational equity, emphasising its moral and ethical dimensions; supporting a framing arising from Thunberg and the wider protest movement itself, where motivational collective action and hope are important framing devices (Hayes and O'Neill 2021, Molder et al 2021).

There can still be issues of equity and representation within these visual portrayals, however; there have been two incidents of Ugandan youth climate activist Vanessa Nakate being cropped from a news photograph where she was originally pictured standing alongside Greta Thunberg, first at Davos in 2020, then at COP26 in Glasgow in 2021 (Evelyn 2020, Hayes and O'Neill 2024).

There is mixed evidence in terms of how people engage with climate protest imagery (again, with the caveat that research to date has focussed on the Global North). An Australian study asking people to select one image they most closely associated with climate change (from a wide sample of representative images) found very few people selected a climate protest image as the most engaging (Leviston, Price and Bishop 2014). The Q-method studies referred to earlier both included a climate protest image: while the UK, Australian and US study found an image depicting a mass street protest promoted a feeling of being able to act on climate change (self-efficacy), it did not raise a sense of issue importance (salience); the German, Austrian and

Swiss study found an image of protesters outside a coal-fired power station invoked a sense of saliency but not self-efficacy (Metag et al 2016). In a German, US and UK study, Chapman et al (2016) found protest images were the climate visuals which attracted most cynicism and could be particularly disengaging, a finding which echoed across the political spectrum and value positions on climate change. Similarly, Hart and Feldman (2016) found a climate march protest image neither increased a sense of saliency or efficacy.

All of these studies focus on images of mass protest. In contrast, there has been very limited attention to how people interpret and engage with performance protest in climate communication. An exception to this is the talk-aloud research of media and journalism scholar, Antal Wozniak. He carried out a study with participants in Germany, asking them to talk through a series of eight performance protest images. His results indicate a substantial disconnect between the intended messages of the environmental activists staging these protests, and the comprehension of these symbolic visuals. Only one of the performance protest visuals was interpreted as intended, with many of the other visuals subject to a wide range of competing and inconsistent interpretations (Wozniak 2021). Future studies into how people engage with climate imagery could usefully examine both mass protest and performance protest visuals; as well as seeking to understand people's attitudes towards social movements as a likely key moderator of attitudes towards climate protest visuals (Kilgo and Mourão 2021).

Opinion shapers online

This book has concentrated on the form and role of visuals as a communicative form. However, it is clear when examining climate visuals of people, and especially with the dominance of online communication through social media platforms, that engaging with multimodal forms of climate discourse is of growing importance. Van Leeuwen (2004) describes multimodal communicative forms as those where there is 'stylistic unity' between the image, typography and layout; and where the verbal and the visual are cohesively and inherently linked. This final section therefore considers two multimodal forms where text and visuals are intrinsically linked to communicate about climate change: in memes, and in Facebook advertising.

The word 'meme' has become 'an integral part of the netizen vernacular', which can be defined as 'the propagation of content such as jokes, rumour, videos, or websites from one person to others via the Internet' (Shifman 2013: 362). While memes may be passed off as an irrelevance, they can play an important role in popular political commentary through their use of humour and satire. For example, memes became a key tool for delegitimising Presidential candidates Donald Trump and Hillary Clinton in the 2016 US elections (Ross and Rivers 2017). In climate debate, memes became a viral

Figure 6.5: 'Condescending Wonka' meme

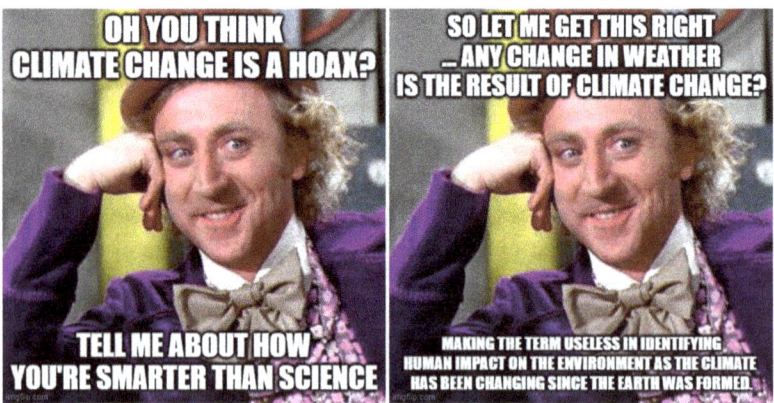

Note: The meme is used to supported two opposing positions on climate change: a 'convinced' logic, and a 'sceptical' logic.

Source: See critique in Ross and Rivers (2019)

tool used effectively by Greenpeace to oppose Shell's plans to drill for oil in the Arctic (Davis, Glantz and Novak 2016); with their effects including an increase in individuals' intentions for civic engagement (Zhang and Pinto 2021).

Linguistics scholars Andrew S. Ross and Damian Rivers examined a set of common meme templates for how they constructed climate change, in terms of their adherence to either a 'convinced' or 'sceptical' logic (that is, what Coan et al 2021 would classify as support or opposition to the 'global warming is not happening' contrarian super-claim; see Ross and Rivers 2019). They discuss an example of a meme template used to support both a 'convinced' and 'sceptical' logic (Figure 6.5). This template, 'Condescending Wonka', features a clip from the 1971 Willy Wonka & the Chocolate Factory film, a movie which has gained cult status in the US and beyond (Library of Congress 2014). This meme template emerged in 2011 and has been described as one of the most popular Internet memes (Nieubuurt 2021). It depicts the film's main protagonist, Willy Wonka, an extremely eccentric, unpredictable chocolate factory owner. Wonka is pictured in a mocking pose: eyebrows raised, smirking smile, head resting ironically on hand in a display of fake listening. The meme text is used to signal a topic in an ironic way, where condescension is used as a rhetorical device to mock an imagined audience (Nieubuurt 2021). The meme's set-up is intentionally divisive, as it posits that there is only one correct or obvious answer (Shenton 2020). The meme's text and imagery are intrinsically linked in setting up the mocking sarcasm which is key to the popularity of this communicative form.

The meme's versatility is such that it can be used to support opposing views. Just as 'Condescending Wonka' has been used to support and oppose US gun control legislation (Shenton 2020), it has been used to support both a 'convinced' (Figure 6.5, left) and 'sceptical' (Figure 6.5, right) position on climate change. In the left side meme, Wonka is used to mock people who are sceptical or climate change as being not intelligent enough to understand climate science (Ross and Rivers 2019). Similarly, the 'sceptical' meme version also mocks unintelligent viewers, but this time, the mocking is directed at any viewer who thinks climate change is happening, for their perceived failure to grasp the concept of natural variation in climate cycles (that is, making use of contrarian sub-claim 2.1; 'human greenhouse gases are not happening … it's natural cycles' Coan et al 2021).

The text plus image form is perhaps the most common type of meme (Ross and Rivers 2019). Images are just as important as the text in these multimodal forms; where often iconic images will be used as a cue or reference to a much more expansive set of ideas. 'Image macros', or non-moving images with superimposed text commonly take the form of a piece of formulaic top text, and a punchline delivered in the text at the bottom (Dancygier and Vandelanotte 2017). These communicative forms are not only evident in viral memes in inter-personal communication online. They are also used strategically by organisations, for example through posting advertisements on Facebook (colloquially known as 'sponsored posts').

Iain Weaver, together with our wider team of computational and social scientists at the University of Exeter, led a piece of research to explore how climate change had been framed in Facebook's repository of adverts relating to social issues, elections or politics (the Facebook Ad Library) from the US, Canada and UK. The image macro was a common communicative form in this dataset. Also prominent was the presence of people in the images: politicians in the US, and a wider range of people in Canada and the UK (for example, activist Greta Thunberg and broadcaster David Attenborough; see Figure 6 in Weaver et al 2022).

Drawing on the work by Weaver et al (2022), computational social scientist Ned Westwood carried out a mixed-methods study of the Facebook Ads Library dataset to explore how memes in these sponsored posts set people up as particular types of character in climate discourse (Westwood 2024). Again, a majority of the memes used the image macro form, and again, were categorised into the 'sceptic' or 'convinced' logic groups do (though note that these terms in quotes can themselves be controversial; see Boykoff and O'Neill 2010, Coan et al 2021). Westwood (2024) describes how 'sceptics' used climate memes to ridicule opponents, characterising their opponents as fools and villains, reflecting the hate-humour nexus (where serious/non-serious cultural expressions of the comic and the political collide, summed up by violent or hateful content presented as 'it's just a joke'; Askanius 2021).

In one meme, a 'convinced' logic is utilised (Westwood 2024, table 10). The top right-hand side of the image depicts a typical piece of political photography featuring President Biden giving a speech. Biden's expression is neutral, if serious, and, in combination with his hands raised up in emphasis, expressive and authoritative. The stars and stripe of the US flag behind him signal power, politics and patriotism. The bottom right-hand part of the image contains the logos for the four highest emitting investor-owned fossil fuel companies. The text in the bottom left hand box delivers the 'punchline' (Dancygier and Vandelanotte 2017), that fossil fuel companies are making excessive profits at the expense of everyday people. There is some nuance here, in that this meme is not supporting the existence of anthropogenic climate change directly, but is instead setting up fossil fuel companies as oppositional actors who are deceitful, greedy and not to be trusted.

Westwood shows another typical meme using a 'sceptical' logic (Westwood 2024, table 10). This meme features the logo of Turning Point USA, a conservative non-profit organisation. The main image shows US Democrat Alexandria Ocasio-Cortez; with her mouth wide open, eyebrows raised and eyes staring, shouting into a microphone. Compared to the typical 'statesmanlike' picture of Biden in the meme described, this picture constructs Ocasio-Cortez as angry, perhaps a little out-of-control: a facial display of anger like this, from a woman politician, is punished by voters (Boussalis et al 2021) – so this is intended as an unflattering portrayal. Ocasio-Cortez is well-known for being a political progressive who has campaigned strongly on climate change, as well as being the youngest women and first ever from the Bronx to serve in Congress. During the time this sponsored post appeared, she had announced the Green New deal, a significant US policy debate on climate and energy. The text of the sponsored post and others like it in the dataset work, even without explicit reference to climate change, to contextually portray Ocasio-Cortez as a socialist or communist threat which is attempting to increase government control and reduce Americans' freedom; and to undermine support for climate action. As for the Biden meme, the text and image work together in an attempt to engage (and persuade) viewers.

Where are the everyday people?

Visuals featuring identifiable people are one of the most common types of images associated with climate change: from newspapers (O'Neill 2013) to TV coverage (León and Erviti 2013), Twitter posts (León, Negredo and Erviti 2022) to TikTok videos (Hautea et al 2021). Yet, a deeper exploration reveals that much of this imagery features particular types of people, particularly political figures. Everyday people in climate visuals – people impacted by weather extremes, for example – are largely missing from the visual discourse on TV (León and Erviti 2013), in newspapers (O'Neill 2013) and online (Hopke and Hestres 2018, Weaver et al 2022). Decisions made by people lie at the heart of a transformation to a sustainable future. Achieving the urgent and necessary transformations laid out in the IPCC reports (IPCC 2023) will require placing people at the heart of climate action (Devine-Wright et al 2022). However, if everyday people are missing from visual representations, and if current visual representations focus narrowly on depicting mostly political figures, this is likely to push people to feeling disengaged with the issue. This final empirical chapter ends, then, with a now familiar call to action, much as the other chapters have concluded: to engage critically and productively with the current visual discourse around climate change (and the opportunities, challenges and barriers these may present in terms of imagining climate change) and to work in creative ways to expand to a more inclusive and engaging visual discourse.

7

Conclusion: The Flow and Friction of Climate Images

The empirical chapters in this book have explored how climate change images are produced, what climate images may look like, how climate images are read, and how climate images move. This conclusion chapter brings together the many different case studies to explore the work and power of climate images.

Climate visuals move through a complex, global media ecosystem

In his seminal work proposing the encoding/decoding model of communication, Stuart Hall described three key moments in the communications cycle: the sites of production, content, and audiencing, all three occurring in a continuous circuit or loop (Hall 1980). This work has underpinned much scholarship since, including on visual communication. However, it is no longer enough, given the fundamental changes in technologies since the 1980s, to focus on reading images as static objects occurring at three distinct moments (Rose 2016); or indeed, to imagine visual news discourse as produced just by select individuals within a newsroom (Seelig 2005, Hayes and O'Neill 2024). The role of global image agencies, of social media platforms, of news wires, of media conglomerates, of the smartphone in the pockets of the majority of the people on the planet (Shanahan and Bahia 2023) – these all work to shape the types of images that we see, including the visuals which come to represent climate change. Rose suggests that focus must, therefore, shift to encompass understanding and mapping of the dynamics of how images move; at digital interfaces (comprising hardware, software and humans; see Law 2002) and across networks (including examining how dispersed actors hold and exert diverse forms of cultural power and agency, Rose 2016).

One way of conceptualising this is to imagine the movement of images as the flow of energy and nutrients through an ecosystem, from photosynthesis to consumption, to predation, to decay. As visuals flow through this ecosystem, they are buffeted by complex interactions between diverse and dispersed organisations, groups and individuals across diverse platforms and their architectures (Anderson 2013, Zuckerman 2021). Thinking through the idea of images and flows also brings to the fore the concept of the Hybrid Media System (Chadwick 2017); and how mutual interactions exist and evolve between different types of media due to their interdependence. An interrogation of images must understand the image in relation to how it was produced and distributed, as well as interpreting the visual and aesthetic cues it uses and how these may have been shaped over time (Aiello 2023). Thus, in order to understand the work images do in the world, it is important to explore how images exist and move in a complex globalised media ecosystem.

Therefore, this conclusion chapter brings together the varied and diverse images and their connections to climate change presented throughout the book, to think through this idea of images moving within a complex and global media ecosystem, with a particular focus on newsmaking. In this way, this conclusion chapter will approach the question: which factors shape the climate visual discourse? And, leading on from this, it seeks to answer those questions as posed in the book's introduction chapter: who has the power to shape the climate visual discourse; and, how might the climate visual discourse be opened up? Using the concept of images moving within a media ecosystem, the following discussion is broadly split into two sections.

The first section, 'organisations, groups, individuals', recognises that actors (broadly understood) can be powerful in shaping visual representations of climate change. This section focusses on news organisations and image agencies as powerful actors that particularly influence the visual climate discourse. It also explores how the imagery we hold in our minds can be powerful in shaping visual climate representations. The second section, 'information architectures', recognises that the ways in which information is organised and presented – for example in both the frontend and backend design of a social media platform – fundamentally influences how that platform is navigated, experienced and engaged with, including for climate change (Freelon 2015, Treen et al 2022). This second section explores ways in which power to shape the visual climate discourse is exerted through ways of organising information; again, synthesising from the case studies of earlier chapters. Of course (as exemplified through the use of the ecosystem concept), there are multiple ways and places in which these two sections overlap and inform each other; but they are intended as a device to structure the discussion in terms of the myriad ways in which power is exerted on the flow of images through the media ecosystem.

Organisations, groups, individuals
News organisations

News media organisations hold considerable power to produce and reproduce meaning, including for shaping the cultural political life of climate change through visual imagery (Beck 1992, Carvalho and Burgess 2005, O'Neill 2013). The case studies in this book have brought to light some of the ways in which particular types of climate visuals may be promoted (or marginalised) through a news organisation.

There are many interacting factors which influence how likely it as that a climate image is selected in the newsroom, from what is often a very large pool of potential options. Some of the landmark scholarship in journalism studies has attempted to define 'news' (Galtung and Ruge 1965, Harcup and O'Neill 2001, Harcup and O'Neill 2016). By identifying 'news values', these studies have all tried to answer the question: which features influence news selection? As O'Neill and Harcup (2009: 162) explain, this scholarship 'goes to the heart of what is included, what is excluded, and why'. The most recent of these studies is Harcup and O'Neill (2016: 1482), which importantly includes consideration of the social media platforms transforming the news ecosystem. They list 15 news values, including bad news (that is, with negative overtones such as death, injury, loss); conflict (controversies, arguments, fights); drama (escapes, accidents, rescues); the power elite (concerning powerful individuals, organisations or institutions); and magnitude (significant in terms of number of people or occurrence of an extreme event). Powerful visuals also feature as a news value in and of themselves; that is, arresting images can drive the news story; as well as forming part of the judgement about the newsworthiness of a potential story (as shown in the underwater cabinet meeting in the Maldives, Figure 6.4). Bednarek and Caple (2012) explicitly consider the role of images and news values, finding that certain photographic devices (such as referencing an emotion) can meet several news values (for example: novelty, negativity and personalisation).

From this discussion of news values, it is clear why, at the level of climate image 'type', certain kinds of climate images might make the news. For example, imagery of climate impacts, such as the people being rescued from a flood in Figure 3.5 were found to be one of the most common types of climate image (O'Neill 2013), perhaps because they meet many of the news values as described. News values may, therefore, help explain why certain types of climate image are more or less common in the news ecosystem. For example, images of a flood have a clear tipping point, a moment of drama – when water spills from its everyday location into places where it is not ordinarily. A photograph which captures the moment that swirling brown floodwaters cover streets and pour through people's front doors satisfies news values of drama, surprise, bad news and magnitude. The flood images by David Maurice Smith (Figure 3.6) though, differ from usual flooding media imagery; his portraits reveal expectation and waiting, reflection and contemplation, to illustrate the long-term impacts that flooding has on human health and wellbeing. These images stand in stark contrast to the high drama of the first moments of a flood event. It is notable that *The Guardian* chose the only image in this series featuring floodwater – a 'typical' flood image – to lead the article about the town of Lismore being flooded once again (see Hinchliffe 2022).

In contrast to flood images, the impacts of a heatwave are much more difficult to visually portray. For example, extreme temperatures can be made visible through their less obvious impacts on human bodies. Moreover, the human bodies most impacted – such as older people – may be less visible, as extreme heat is often endured in the private spaces of people's homes rather than the public space of the streetscape. Also for a heatwave, there is a less obvious threshold of perceived risk. Whereas floodwater pouring onto a street is a clear indicator of something 'out of place', there is nothing so obvious for a heatwave. Should this threshold be a particular temperature point on the thermometer, perhaps, as in Figure 2.2? Even this, however, is culturally mutable, as a 'heatwave' is defined differently by meteorological organisations in different parts of the world depending on prevailing climatic conditions and societal norms. Thus, the news value of 'magnitude' is harder to visually portray for some climate images than others.

How a story exists or moves across news desks or beats may also influence the images which come to exist in the climate visual discourse. A beat or desk is the thematic area that journalists specialise in, for example, the politics desk, the business and economics desk, the health desk or the science and environment desk. Decision-making and preferences about image choice may change depending on the specialist knowledge which is held by journalists, editors and other newsroom workers as a story crosses different news beat.

For example, Chapter 6 discusses a photograph taken at COP26 in Glasgow, of Scotland's First Minister, Nicola Sturgeon, standing with youth climate activists Greta Thunberg and Vanessa Nakate. Although this image went out on the wire with both Thunberg and Nakate pictured, the final image appeared cropped – removing Nakate – in at least four online news articles. This cropping was especially egregious given that it was not the first time that Nakate had been removed from a photograph; despite her key role as a youth activist from Uganda and a prominent actor raising up the concerns of the Global South (see Evelyn 2020). Responding to the image crop, Nakate commented that the news outlet 'didn't just erase a photo, you erased a continent' (Nakate, in UNICEF 2024). Research interviews with journalists (not associated with the crop) suggest that how the story crossed news beats could be responsible for the error, with general news reporters perhaps unaware of Nakate's important role in the climate negotiations, unlike climate and environment specialists (Hayes 2024, Hayes and O'Neill 2024, see also Strauss et al 2021).

Similarly, as discussed in the context of heatwave imagery in Chapter 2, the dominance of 'fun in the sun' images may also be partly influenced by the reporting of the news story by general news reporters rather than climate specialist journalists. The nuance of ensuring visual reporting matches the risk portrayed in news article text held by specialist environment reporters may be lost, as other factors (for example, aesthetic choices about the value of colour for the layout of the newspaper page, or perceived viewer preferences) may be valued more highly by staff with different expertise as the heatwave story moves to a different part of the newsroom. This observation is not unique to climate reporting; Journalism scholar Kimberly Bissell described how 'it was routine for non-photo staff to make decisions about photographs [at a later stage in the news story process] … selection was made according to what was appealing rather than on how well the photograph told the story' (Bissell 2000a: 11). Along similar lines, different types of news visuals for ostensibly the same story may arise between different types of news organisations. Visual communications scholar TJ Thomson argues that local news journalists often have to make images or videos alongside their other duties including writing, editing and social media management. This lack of visual specialisation and formal training, plus time pressures, means that their images are more likely to perform a storytelling function (Thomson 2024).

Another factor influencing image choice is concerned with the nature of the newsroom as a fast-paced environment. Many journalists emphasise how image searches for their stories can be laborious, and that oftentimes, they need to be quick in selecting the first image that fits a story. Additionally, there are many people involved in image selection in news organisations, constantly making quick decisions about images (see also Bissell 2000b): 'microjudgements are made all day, everyday, by many, many people, in a collaborative, fast-paced environment. These decisions go on to determine the visual news coverage' (Alastair Johnstone, Deputy News Picture Editor, The Times, 2018–2022, personal communication, 5 January 2024).

As Chapter 4 explores in the context of climate energy imagery, journalists often need to select images at considerable pace to fit with the intense workflow of the newsroom. The need for speed can, therefore, be an important factor shaping the visual discourse of climate change, then; with images which are front-of-mind and easily accessible (both in terms of people's cognitive schemas in terms of what comprises a 'climate' image; and in terms of being easily discoverable) more likely to be selected.

A further factor shaping image selection may be how the image is to be featured relative to other modalities such as text or narration, as well as other types of content such as advertisements (De Smaele, Geenen and De Cock 2017). For example, in a newspaper, most images directly accompany a piece of text-based news, to provide more context on that story. In these 'news pictures', the content and the context of the image are considered in detail, alongside aesthetic choices. However, 'standalone' images also exist. A standalone newspaper picture is an image and caption that exists as a mini-story in its own right. The image is not connected to any other longer-form text or content on the page (Reque et al 2001). Although there are fewer standalones, as their selection depends on other factors such the news environment of the day ('how much' news there is) and the amount and placement of advertising, standalones are intended to be compelling pictures in their own right. Standalones are intended to inform or entertain viewers without the need for lengthy sections of text (Whitmore 2018). The aesthetic value of a standalone picture, therefore, becomes more important than for a news picture. In terms of climate images in the news, then, standalone pictures may be one reason for the proliferation of certain types of images – such as the 'fun in the sun' heatwave visuals (see Figure 2.1). As a quote from Greg Whitmore, *The Observer* newspaper Picture Editor makes clear, a standalone picture is 'usually a photograph of a young woman at a music festival or a crowded beach on the south coast' (Whitmore 2018: np). A 'fun in the sun' visual can present the opportunity to insert a standalone picture into the news; a bright, saturated, colourful image which balances the dark text on the rest of the page:

> There is a tendency for heatwave images to be considered as 'standalone' images (leading to more aesthetics-based decision making) rather than as 'news' images where the content and context are considered in more detail alongside the aesthetic choices. I believe this leads to the role of these images being restricted to an aesthetic contribution to the page, rather than as adding visual information or context to the story they accompany. (Alastair Johnstone, Deputy News Picture Editor at *The Times*, 2018–2022)

Image agencies

As early as 2003, scholar of visual culture, Paul Frosh, raised questions about the power of global image agencies, and particularly their role as the purveyors of the 'wallpaper of consumer culture' through provision of generic images (Frosh 2003: 1). Frosh defined a generic image as 'referentially ambivalent', or a staged performance using paid models to represent widely recognised image-genres (happy family on holiday, gay couple at a restaurant and so on). Its purpose is to be sold and re-sold in a diverse range of situations for maximum commercial gain; with many of these future roles unanticipated by the photographer or image agency. Aiello (2023) compellingly describes a case study demonstrating exactly this: Getty photographer Jeff Mitchell sympathetically shot an image of migrants crossing the border between Croatia and Slovenia. This image was licensed not just for editorial purposes, but also for commercial and promotional purposes. It was completely decontextualised, in combination with the 'BREAKING POINT' red text overlaid on it, in a political poster campaigning in the Brexit referendum by UKIP politician Nigel Farage to instead present a racist visual narrative. Although Frosh refers to generic images through a study of stock imagery, he also discusses how the concept can also be applied to photojournalism, in that there are a number of similar, predictable and limited image types of photojournalism; providing such examples as 'mourning woman' or 'a distressed witness to an atrocity' (Frosh 2020: 20).

A few years later, scholarly work exploring the power of image agencies specifically began to critique the role and power of global images agencies in 'visually branding the environment' (Hansen and Machin 2008: 777). Yet, there are a limited number of content analysis studies of climate imagery; with fewer still of these documenting the provenance of images. How prevalent are image agency visuals within climate news? The studies which do report image copyright data show the huge reach and influence of global image agencies such as Getty Images, Agence France-Press, Associated Press and Reuters. These image agencies variously supply either or both of 'creative' or stock images (for marketing and corporate use, though they do also feature in news coverage) and 'editorial' images (which can only be used for informational or descriptive purposes in newsworthy contexts). In work led by Geographer Sylvia Hayes and discussed in Chapter 6, we have found Getty Images dominate climate reporting, at least in the climate protest and climate political summit visuals analysed to date. In that 2019 UK newspaper dataset of almost 750 climate protest images, images were overwhelmingly credited to image collections; with almost a quarter (23 per cent) of all the images credited to Getty Images or one of its subsidiaries (Hayes and O'Neill 2021). Similarly, in research examining the coverage of COP26 in Glasgow, most images were attributed to an image agency, with over a third (35 per cent) of all images attributed to Getty Images (Hayes and O'Neill 2024).

The visuals discussed in Chapter 2, in the context of heatwave risk, are especially interesting to consider in this discussion of generic imagery. The thermometers, blazing sun and starbursts of the heatwave imagery discussion (see Figure 2.2) are clear examples of stock imagery, which are used across a range of circumstances and media types to communicate the 'idea of heat' (O'Neill et al 2023). Figure 2.3 shows a photograph of a Black man wiping his brow, visibly and uncomfortably hot, sitting at the wheel of his delivery van. It is an example of a heatwave image which does communicate something of health risks and vulnerability to an extreme heat event. However, note how it was used in BBC coverage in the UK, but the image is of a US-based driver (the steering wheel is on the opposite side in UK vehicles). This image demonstrates the globally networked ecosystem of images. Even where images are not 'stock' photographs, they often still meet the criteria of Frosh's generic image; such as in the happy beachgoers, colourful parasols, sea and sand of the 'fun in the sun' heatwave imagery (see Figure 2.1, and O'Neill et al 2023: 90).

The images in people's minds

The images which exist in people's minds about climate change (also called 'affective imagery', see Leiserowitz 2006) can also be powerful in influencing the sorts of images which are quickly brought to mind, produced and shared. As the introduction chapter discusses, individual frames are clusters of ideas in people's minds which guide how they process information (Entman 1993). Individual frames (in the minds of journalists, for example) feed back into the media frames which come to exist (Engesser and Brüggeman 2015).

This is evident through several of the case studies discussed throughout this book. Particular image types start to become frequently used to represent climate change. As they are used more, they can become iconic, rooted in a feedback loop, and embedded in the visual discourse. For example, 'fun in the sun' images (Figure 2.1) are a front-of-mind visual for photojournalists tasked to report on heat extremes. As a result, when a heatwave occurs, the response is to 'hop on a train to Brighton' (a UK coastal resort close to London; as was memorably explained to me by a picture editor) and produce even more of this same type of imagery. An image of a young Melanesian child walking through a shallow flood is selected by photo editors to accompany numerous topically different climate stories at one newspaper outlet (Figure 2.5). Consider also how the smokestacks image (Figure 4.5) is quickly brought to mind and hastily inserted as a suitable climate image by journalist working under time pressure. Likewise, polar bear images have evolved from being simply associated with news stories about sea ice decline, to being associated with news and other content which makes no reference to the polar bear picture, or even climate change – the cultural association between 'polar bear' and 'climate change' is considered so strong as to need no explanation (see Chapter 3).

Such images have come to be visual metonyms of climate change. They have become a type of visual shorthand (used within a particular culture) which goes beyond the immediately represented denotative content to directly link to a set of ideas about climate change in real or conceptual space. In turn, the visual metonym itself can then become indistinguishable from the (often deeply political) judgements with which these ideas are associated. It comes to stand in for a set of unspoken assumptions about climate change (O'Neill 2022: 3; see also Scarles 2004, Perlmutter 2006, Aiello and Parry 2020). And, as Linder (2006) described in his study of climate change advertising which uses the closely aligned rhetorical device of synecdoche, such images may also become subject to parody, inverting and challenging the original reading of the visual (exemplified by the polar bear cartoon; Figure 3.2). At many points in the visual media ecosystem – the photojournalist, the image agency, the newsroom, the person who reads the news image (and many others besides) – such images are easily brought to mind and expected. These image types are sought out, constructed, captured, uploaded, stored, selected, viewed, shared, remembered (see also Frosh 2020). While such image types simultaneously easily connote something (viewers can easily locate the image as a 'climate image'), at the same time, such images can come to mean very little at all as they 'slide towards radical interchangeability' (McQuire 1997: 59).

This discussion provides an indication of the dominance of powerful actors in shaping the visual discourse of climate change. It also shows how the loss of newsroom or staff photographers associated with particular news organisations (Gynnild 2017) can impact the visual discourse; with the increasing reliance on image agencies having implications for the types of visuals that are more accessible for news organisations to use in climate news coverage: likely more generic, more abstract images – in order that the licensed images can be used across as wide a range of contexts as possible for commercial advantage (Hansen and Machin 2008). In becoming more generic though, these images can be more likely to reinforce cliches or stereotypes, or may act to 'other' the human subjects featured (Hansen and Machin 2008, Aiello 2023). This is not to say that gatekeeping decisions made by photo editors and others are not important – far from it, their decisions play a key role in shaping the visual frames used to report news (Fahmy et al 2007). Instead, this is a call to recognise that image agencies hold considerable power in normalising certain types of visuals over others. The increasing dominance of image agencies is occurring at the same time as the ongoing decline in new staffer photography jobs (Hadland and Barnett 2018). Taken together, these factors can lead to the homogenisation of images and imaginaries (Davoudi and Machen 2022). This raises a significant challenge to a conclusion returned in many of the empirical chapters of this book; how to more effectively feature 'everyday' people in climate stories.

Information architecture

This book has detailed many examples of how particular visual tropes are repeated time and again across different media. There are also examples, though, of not just the same image type, but exactly the same image, being repeatedly used. I found the UK's *Daily Express* newspaper reused the same photograph of a calving ice shelf eight times in just one year (meaning around 7 per cent of the *Daily Express* climate visual coverage in 2010 used this same photograph; O'Neill 2013). Chapter 2 explores multiple reuses of the same photograph in *The Guardian* newspaper; an image featuring Miriam Bulivono, a young child walking alone through a flood in Matacawalevu, Fiji. It was used at least seven times between 2017 and 2023 to illustrate a diverse set of news stories, as well as two opinion pieces. The caption, posted next to the image on a stock image website, mentions an extreme weather event – Cyclone Winston – but the cyclone happened almost a year before the photo was actually taken. Similarly to the aerial image of a coral atoll (Figure 2.4), the photograph of Miriam Bulivono demonstrates how Pacific island imagery travels through a globalised, highly-networked media ecosystem to become a climate visual metonym (O'Neill 2022). This reflects findings of Spiegel's (2020) study of visual storytelling of mining in Indonesia, where a story about Indonesian mining was illustrated with a stock image of mining in Peru. As these examples have highlighted, images attached to news about a place can be conspicuously detached from the realities, meanings and lives of people in that place; they are an abstraction.

Although the seven uses of the image by *The Guardian* newspaper over the six-year period are broadly about climate change, the image captions and article texts often do not mention Fiji or even the wider Pacific region; demonstrating how a photograph can come to stand in for a set of unspoken assumptions about small island vulnerability; and about power and responsibility for climate action. Climate scientist Simon Donner, in a piece critiquing a Google Images search for 'victims of climate change' argues similarly; that the dominance of such portrayals

> reflect the assumption, common among individuals in the global north … that people in tropical climates are helpless victims who lack the capacity to cope with climate change. Beneath that assumption lies a long, ugly history of climate determinism: the racially motivated notion that the climate influences human intelligence and societal development. (Donner 2020: np)

This next section now explores how information architectures – that is, the ways in which information is organised (both about and within images) – can be powerful in marginalising or promoting particular ways of visualising climate change.

Managing, sorting, searching

There are a huge number of visual images available, via, for example, image agencies: the Getty Images search analysed in Chapter 4 returned more than

10,000 images in a search for 'wind turbine' images (see Figure 4.4). There is also an enormous daily throughflow of images through to newsrooms. *The Guardian*, for example, had an archive of over 3 million images in 2015, with around 20,000 new images added per day at that point (Cevey 2015). Why then, might organisations repeatedly use the same visual?

Given the enormous quantity of images circulating in the visual media ecosystem, organisational structures have necessarily been developed so that news organisations can manage – access, sort and search – this vast quantity of information. Chapter 2 discusses one potential explanation, that of the management of visual content within the news organisation. Organisational structures such as *The Guardian* Grid (image management system) are designed to assist in accessing, sorting and organising visual content (Cevey 2015, Hayes and O'Neill 2024). Ways in which these systems allow filtering by criteria such as cost (for example, preferentially returning images in a search for which an organisation already has an image-rights subscription); or due to their being already archived in a content management system; raise the chances of an image being reused.

Another important consideration, in terms of image selection, is an understanding of how an image search is carried out. One way of selecting an image from a searchable collection is through text-based queries, which search the text-based metadata assigned to each image in the collection. The metadata – including title, description and tags – can be literal keywords which describe a person or an object: 'President Trump' or 'wind turbine', perhaps. The labels can also be more ideological or conceptual, describing socio-cultural or personal associations, including emotions: 'climate delay', perhaps a tag alongside the 'Trump' tagged image, or 'hopeful' alongside the 'wind turbine' tagged image (see also Panofsky 1970, Hall 1973, Dyer 1982). For example, a typical wind turbine image (see also Figure 4.3) on Getty Images (ljubaphoto #1316576079; Getty Images 2024a) uses the tags 'Wind Turbine', 'Renewable Energy' and 'Environmental Issues' but also 'Awe', 'Idyllic' and 'Aspirations'. Tags may be added at different stages, including by the image creator or photographer, or later on by image agency staff (Getty Images 2024b).

These search options help to narrow down a vast pool of potential images to fewer, more relevant options. However, journalists and others working in the media ecosystem report considerable limitations in searching for climate images for their work. This is well explained by a climate journalist and editor, next:

> I've often had issues finding good climate images. Climate-tagged photos tend to be very heavy on COP. Often, photos from climate-related floods, droughts, or heat aren't tagged as such, making them quite hard to find. Climate-related photos that aren't tagged to a specific disaster, and ones that are taken in smaller countries and dated from

within a couple years, are often non-existent. The result is that if you look closely, you will actually see the same climate-related photos pop up again and again, all around the world, year after year. (Katherine Dunn, Content Editor at the Oxford Climate Journalism Network [OCJN])

This quote elucidates some of the clear limits encountered while searching for climate images. Most obviously, many potential climate images are not tagged as such in image databases. At present, tagging images as 'climate' images appears to be quite limited – to political events such as the annual COPs. As Chapter 6 explores however, the COPs mainly bring about a consistent and 'dreary' visual discourse mainly featuring politicians (Eide 2012, Grittmann 2014, Wozniak et al 2017: 1436, Hayes and O'Neill, in review, see Figure 6.3); images which largely fail to engage audiences on the topic of climate change (O'Neill et al 2013).

The labelling of an image can be clearly on display or otherwise easy to access. For example, on the Shutterstock stock imagery website, images are openly labelled with 7–50 keywords (Shutterstock 2024). In other cases, though, labelling of images comprising a search may be less clear. In a study of Google Images, Pearce and Gaetano (2021) found that images appearing in Google Image search results for climate change appeared to be ranked on whether they fitted Google's visual vernacular of climate change as a time-less, place-less, human-less and cause-less issue. As with any content, tagging has the potential to produce racialised, gendered, stereotyped or otherwise problematic labelling of images.

Instead of searching via an image's text-based metadata, an image collection can instead be searched by value of its visual properties (such as colours, shapes or textures) via computer vision techniques, in an approach called a 'reverse image search'. In this alternative approach, no metadata (keywords, tags) associated with the image is needed to carry out the search. However, a reverse image search does rely on having an initial, suitable image on which to build a search query, as well as a powerful and competent search facility. At present, while reverse image searches can be useful (for example, to undertake fact-checking for visual misinformation, AFP 2020), they appear to be of limited utility for opening up the visual discourse or discovering alternative climate visuals. Katherine Dunn, the journalist and editor quoted before, recounted that reverse image searches are yet to be widely used to find climate images, at least in the domain of climate journalism.

Culture of the click

There have been tremendous changes in the world of newsmaking and journalism, as the industry adapts to technological changes to create, consolidate or enhance digital content and platforms. In markets across the world, there has been substantial news industry cost-cutting and journalistic layoffs. There is a clear trend for social media becoming more important as an entry point to accessing news stories, rather than readers accessing news directly from the platforms hosted by news organisations themselves (Newman et al 2023). Noting these trends is important, because the presence of compelling visuals in a news story can help to draw in audiences and make a story more shareable (Graber 1990, Thomson and Greenwood 2017, Jaakonmäki, Müller and vom Brocke 2017), a substantial advantage in an increasingly competitive marketplace.

So, compelling visuals can increase shareability online, and hence increase viewers (and revenue). This is not just the case for news organisations: indeed, social media posts containing an image perform best for engaging audiences (Adobe 2014) and, therefore, adding an image is one of the first tips that content creators suggest users should undertake to increase engagement with their social media content (for example, Tamble 2019). As a study of climate protest visuals on Twitter (now known as X) found, posts with images or videos tended to attract more likes and retweets than text-only content; with images featuring people or crowds particularly likely to be widely shared (Lu and Peng 2024). This is all part of what Anderson describes as 'the culture of the click' (Anderson 2013: 98).

The shareability of an image is an important part of some climate visuals' rise to prominence. The Climate Stripes began life as a graphic to illustrate climate scientist Ed Hawkins' public talk at a literary festival, and have since become an online phenomenon via Hawkins' Twitter/X feed, boosted each year by the summer solstice #ShowYourStripes media campaign (see Chapter 6). The impact of this scientific visual going viral has rippled out widely, beyond digital platforms; to feature in public and private life – projected onto the facades of prominent buildings, featuring on professional sports team kits and onto racing cars, even crafted into people's sewing projects and into tattoo art, among many other examples. That the visual is aesthetically appealing, personalisable (the website allows users to create and download a Stripes visual based on climate data from many different locations worldwide), while still being recognisable, and published under a Creative Commons license has presumably all contributed to its success as an exceptionally shareable climate visual.

In a newsroom environment, thumbnail pictures are used on a news organisation's website as part of strategy to draw audience traffic to a story; and thus form an important part of understanding climate visual selection. The format limitations of the thumbnail image – namely, being a very small image which is quickly scrolled past – presents a challenge for newsrooms. An image needs to be very clear at a small scale and draw people into a story quickly. This may mean a greater reliance on cliched images, such as a bright, colourful beach umbrella resting on a sandy beach, as an image thumbnail to represent a heatwave story (see Chapter 2). A BBC journalist explains: 'The overarching priorities for editors choosing images is finding something really simple that works both online and in a small thumbnail on a mobile phone. That rules out many more complex or detailed images' (Esyllt Carr, BBC News journalist, personal communication, 5 October 2023).

Newsrooms increasingly rely on a 'datafied version of the audience' (Dodds et al 2023) with audience analytics now ubiquitous across most news organisations (Petre 2015) (although they are far from the only influence on newsmaking: for example, research with Swedish news organisations indicates that such metrics are important, but only ' as one parameter among many others'; Karlsson and Clerwall, 2013: 70). Individual social media users are also able to easily access metrics on how audiences are engaging with their social media platform content, via free in-app software (an example is the 'post engagement' information provided by Twitter/X); as well as through more specialist analytics services. If audience analytics become a significant driver of content production, it may lead to certain types of images being used compared to others (and it is worth noting here that Karlsson and Clerwall (2013) found metrics played more of a role at some media outlets than others, depending on their nature as either commercial or public service media). For example, a 25-country study by Vlasceanu et al (2024) tested different types of communication intervention on climate mitigation outcomes. They found that the willingness of participants to share information about climate change on social media was increased most by using 'doom and gloom' styled messaging.

The link between doom and gloom messaging and looking to audience metrics as a factor in content production likely favours certain types of visual coverage. For example, this would favour images like the 'emaciated bear' polar bear image (see Chapter 3); which, indeed, became an online phenomenon, reaching an estimated 2.5 billion viewers (Mittermeier 2018). Similarly, the weather map (Figure 5.5) in Chapter 5 which 'breached' (Dixon 2023: 103) the norms of human experience and weather forecasting convention in their new purple colour chart extension also quickly went viral, becoming a defining moment in Australian history, resulting in it being archived in the National Museum of Australia (2023; see also AMOS 2023). It is important to note, though, that while the Vlasceanu et al (2024) study showed that doom and gloom images might increase clicks and shares, such images seem to have no effect on people's beliefs in climate change or their support for climate policies; and in fact appear to significantly lower people's likelihood to take personal action on climate change. Their finding echoed one of our early studies on climate image audience reception, which suggested that 'fear won't do it' – or more precisely, that images which invoked fear (like the smokestacks in Figure 4.1) tended to also lead people to feel helpless and overwhelmed to act on climate change (O'Neill and Nicholson-Cole 2009). As a result, relying on images which increase shareability and (at a surface level, at least) audience engagement, may in fact lead to greater apathy and alienation from the issue of climate change, and be counterproductive, if a sense of efficacy (being able and empowered to take action) is not also engendered.

Platform and format

The platforms on which a story might be featured may also impact the choice of image selected, to fit a platform format or aesthetic. In 2020, the *Guardian Australia* website published a news photograph picturing a man lying dead on a street in Wuhan, guarded by medical workers in protective suits. The article used the image in context and, especially as it featured a fatality, the decision to use this particular photograph was one especially carefully considered by the editor (see also Kratzer and Kratzer 2003 and Fahmy 2005 for further discussion of editorial decision-making around publishing disturbing images). However, the role of social media platforms in newsmaking unwittingly changed how the image was viewed, as it transferred from the newspaper's website and onto other platforms beyond the newspaper's immediate control, with their different platform architectures. The image began to be featured on *The Guardian*'s Facebook feed as a standalone image, 'unavoidably detached from the contextualising article' and given prominence based on Facebook platform metrics, rather than a decision made by newsroom editors (Ribbans 2020: np). This situation generated around 800 comments, which were all largely debating the ethics of publishing this particular photo. As *The Guardian*'s readers' editor, Elisabeth Ribbans, reflected: 'it is a reminder that the question of not only "if" to publish but "how" has many and increasing layers' (Ribbans 2020).

In terms of climate images, newsrooms and other organisations working across multiple platforms will need to take into account how images work both contextualised with accompanying explanatory text, but also if they were to be viewed without this textual account. Can they effectively stand alone? And are they appropriate without the wider context provided by the text? An additional, related issue arises in considering image-text congruence, that is, how well matched the text and visual are in meaning. The UN's tweet text about climate risks, vulnerable communities and two million deaths, overlaid on a photograph of two young boys giggling in a deep flood in Bangladesh (see discussion in Chapter 3; UNGeneva 2023) is an example of when image-text congruence can be problematic and even offensive. The memes discussed in Chapter 6 (for example, Figure 6.5) further demonstrate that the same image can be used to communicate opposing positions on climate change ('convinced' or 'sceptical' logics), depending on the positioning text used.

DiFrancesco and Young (2011) were the first to note how climate news stories often pulled in different narrative directions through visual and textual accounts; and it is clear this is still very much the case. Looking again at news reporting of heatwaves, for example, headlines about fatalities due to extreme heat exposure are juxtaposed against 'fun in the sun' images (Figure 2.1, and O'Neill et al 2022). Image-text congruence can be more subtle, but equally important to recognise, in terms of image processing as well as denotative content; bring to mind the composite polar bear image (Figure 3.4) that was clearly inappropriate and counterproductive when juxtaposed alongside a letter from scientists asserting the integrity of climate science, in the journal *Science* (see Gleick et al 2010 and Chapter 3), even though its aesthetic qualities conformed to expected norms of climate change/polar bear imagery.

Science and Technology Studies scholar Warren Pearce and colleagues have shown how climate change is visually represented in perceptively different ways depending on the online platform used (Pearce et al 2018). They illustrate their findings with a selection of composite images 'stacked' on top of each other, a rich and satisfying way of natively portraying visual information in visual form to explain visual aesthetic vernaculars, rather than reducing visual information to text. Climate change information on Instagram is associated with beautiful, stylish photography; the Twitter stack demonstrates the multimodal overlaid text captions over images style; while Reddit is dominated by images featuring political symbols such as flags, podiums and men in suits. This may result in choices being made to ensure a climate image sits within an expected platform aesthetics: a dreamy, ethereal, soft-focus photograph fits the aesthetic on Instagram, but may look out of place on Reddit or on a news organisation's webpage.

The colour of climate

Image aesthetics also play a role in the power of images and how they travel through the media ecosystem. Focussing on the role of colour in climate visuals is illuminating. Here, colour is understood as a function of hue, saturation and value (see Kress and van Leeuwen 2020). Chapter 5 begins by discussing the 'Burning Embers' diagram (Figure 5.1), a prominent (and contentious) IPCC bar chart designed to visually delineate what might be considered 'dangerous anthropogenic interference with the climate system'; a critical part of earlier UNFCCC climate negotiations (Schneider 2001, Dessai et al 2004). Geographer Martin Mahony explores the detailed negotiations between scientists wishing to provide a useful and yet truthful image through the Burning Embers figure. Decisions about colour choice and shading, for example, were robustly debated (Mahony 2015). While the colour blue was suggested by some scientists as a way of showing some potential positive effects from low levels of climate warming, 'heated debates' about whether this accurately represented the state of the scientific literature and how this might be interpreted, eventually led to its removal and an entirely red-hued colour spectrum being used instead (Mahony 2015: 7).

Chapter 5 then discusses Figure 5.5, the Australian weather map which 'breached' both experience (representing a historically unreached temperature record) and convention (by bringing in a new colour to represent this new temperature band, purple); and also the new red and black hues on a UK weather map to indicate extreme heat (Figure 5.6). Again, the perceived truthfulness of these images was central to how they were interpreted (and indeed, how they gathered momentum and visibility). As they moved between media organisations and across different platforms, some called them hoaxes, while others challenged these claims as climate misinformation (see GB News 2023 and Silva 2023, respectively). The BBC's climate disinformation reporter explains how these attacks have substantial impacts on individuals, as well as the wider public discourse of climate change:

> As part of my job, I have witnessed how imagery can be used to great effect by those spreading misinformation about climate change. ... In 2022, I reported how some social media users dismissed record-breaking temperatures across the UK, by comparing them to those registered in the summer of 1976. So, it was no surprise to me that, a year later, when an intense heatwave hit southern Europe, variations of those claims and narratives found their way onto the public debate again. And yet, baseless as they may be, the rapid spread of such claims online has had a tangible impact on weather forecasters, many of whom have reported increases in online abuse and harassment. (Marco Silva, senior journalist, BBC News and the BBC's climate disinformation reporter, personal communication, 8 May 2024)

A critical approach to climate visual communication

The previous section synthesised examples throughout the book to speak to the question: which factors shape the climate visual discourse? Given this discussion, and the current narrow visual discourse of climate change, further questions inevitably arise: who has the power to shape the climate visual discourse; and how might the climate visual discourse be opened up?

As stated in the introduction, it is important to note here that the intention is not to condemn any particular images, their makers or readers. As this book has elucidated, images exist and move through a globalised media ecosystem, with complex and dynamic interactions between images' creation, circulation and (re)interpretation. However, it is important to recognise the limits of the current climate visual discourse; and to acknowledge that there are sites and actors where opportunities exist to open up the climate visual discourse. This section provides some suggestions for interventions that may exert power to open up the climate visual discourse to be more diverse, equitable, inclusive and responsible.

This discussion expands upon how media organisations and image agencies hold particular power to shape and reshape the visual discourse of climate change. Commendable approaches are already looking to influence the ways in which media organisations visually portray climate change, through education and training of photographers, journalists, editors and others at media organisations on visual climate communication.

One such approach is the Climate Visuals programme, run by the climate communications charity Climate Outreach (Climate Outreach 2024). Climate Visuals works with journalists, picture editors and photographers, as well as people in government and civil society organisations, to offer training on climate visual communication. Most notably, Climate Visuals worked with *The Guardian* newspaper to think through the organisation's climate communication visual strategy. This resulted in a contribution to *The Guardian*'s Climate Pledge 2019 and the strategy outlined in the piece: 'Why we're rethinking the images we use for our climate journalism' (Shields 2019):

> as picture editors and photographers, we are having to think again about finding the right focus. Many of the impacts to communities, biodiversity, agriculture, water and food supply represent the escalating crisis our planet faces, yet visually they can be far more challenging to depict. We need new imagery for new narratives. This can be challenging in a fast-paced newsroom but it is important to be nuanced and creative with search terms to unearth photography beyond the usual keywords of climate change, heatwave and floods. (Fiona Shields, *The Guardian* picture editor, quoted in Shields 2019: np)

Similar programmes around the world also offer broader training to journalists on covering climate change. For example, the Oxford Climate Journalism Network (OCJN), part of the Reuters Institute for the Study of Journalism at the University of Oxford, is a training and networking-based initiative. Practising journalists and editors can sign up to an online, six-month course covering many aspects of climate journalism, including visual climate communication. Expertise is shared between speakers and Fellows, but also between Fellows themselves. At a seminar I gave for the OCJN on visual climate communication, an OCJN Fellow shared with others how she had bought about change in heatwave visuals at the media organisation she had worked for: 'In my newspaper, what really made a difference was talking directly with editors and explaining for example the problem in showing pictures of people eating ice cream in an article about the risk of heatwaves. It was only after those conversations that something changed, including the photo editor approach' (Vera Moutinho, multimedia journalist, *Público* newspaper, Portugal from 2013–2022, personal communication, 6 October 2023).

Such training programmes may also cover solutions journalism, defined as investigating and explaining, in a critical and clear way, how people are trying to solve widely shared problems – that is, considering not just the problem, but also potential responses to that problem as newsworthy in themselves (Usery 2022, Solutions Journalism Network 2024). This is relevant to the example discussed in Chapter 3 around photographer David Maurice Smith's flood images series, and the challenges faced in reporting on episodic extreme weather events, which can gather a lot of coverage, but then drop off the news agenda without exploring underlying issues more deeply (see Buckland 2023 for the story referred to in the quote):

> The interest in the [flooding] story completely dropped off. So, I reached out to an editor that I work with a lot … I started making pictures. I was trying to be effective, working with people in crisis, being creative. You question yourself: are we creating entertainment or creating news? There are barriers to making these kinds of stories. So, I talked a lot with this editor, over a long period of time – certainly a long period of time for journalists, about the impact of these floods. One and a half years after the first floods, the [second] story came out. (David Maurice-Smith, photographer, NSW, Australia, personal communication, 22 April 2024)

The Reuters Digital News report suggests that, globally, almost half of people who say they avoid the news would still be interested in solutions journalism (46 per cent of selected markets, 22,467 respondents; see Newman et al 2023: 24). A systematic review of solutions journalism literature clearly shows that solutions news stories have a positive effect on readers' emotions (Lough and McIntyre 2023). Solutions journalism can also improve audiences' mental wellbeing and engagement through psychological empowerment, which is particularly important in the context of issues which require civic engagement (Zhao et al 2022). These studies suggest, then, substantial support for climate solutions journalism. It is worth noting though that there has been some pushback to solutions journalism from climate journalism specialists (for example, see Edwards 2022); and that journalism resources clearly state now that 'reporting on solutions does not mean downplaying the dangers, sugar-coating the facts, or urging a particular outcome' (Covering Climate Now 2023: np). In terms of climate visuals, as multimedia journalist

Vera Moutinho explained further in the context of reporting on heatwaves, conversations about covering climate reporting differently can go hand-in-hand with thinking differently with climate visuals: 'And of course by pitching stories that are more realistic in terms of the heatwave's impacts: not going to talk to people or photograph at the beach, but going to a poor neighbourhood or talking to construction workers ... those stories bring different visuals' (Vera Moutinho, multimedia journalist, *Público* newspaper, Portugal from 2013–2022).

This quote of Vera Moutinho's on newsroom practice is further reinforced via nascent scholarly evidence suggesting that solutions visuals may be increasing. A study of *Público* newspaper climate visuals across two periods (2000–2005, 2020–2022) found that visuals depicting potential solutions and adaptation strategies have increased substantially, by 16.3 per cent between the two time frames (Lopes and Azevedo 2023).

A second approach is to offer newsrooms easy-access, rights-managed photography that works to expand the visual discourse of climate change beyond the usual suspects. For example, the Climate Visuals programme also provides a free library of rights-ready climate photography (Climate Outreach 2024). Images are selected for inclusion based on robust social science evidence about climate communication and public engagement. Similar approaches also exist for film footage. The Capturing Climate Change project is a library of CC BY licensed visuals of climate change from across the globe (Verchot and Biswas 2024). A still larger library of footage free for educational, environmental or impact-led storytelling is available through the Community Interest Company Open Planet. Open Planet was created by award-winning filmmakers Studio Silverback (Open Planet 2024). Open Planet are working to expand climate imagery in film footage to include, for example, imagery of climate solutions, as well as climate impacts:

> Ever since the first human civilisations, storytelling has helped make sense of our world. But today it is not just the story that is important - it is who gets to tell the story. For too long, the visual materials to show the stories of climate impacts and solutions were unavailable to most people: they were expensive, or behind paywalls or perhaps simply didn't exist at all. Open Planet exists to change that: to democratize the images of our changing planet so that anyone, anywhere, can access them free of charge and produce the stories they want to tell. (Colin Butfield, Executive Director, Studio Silverback and Director, Open Planet)

Both visual climate communication training and creating bespoke visual libraries may be powerful approaches to reshape the visual discourse of climate change. At present, however, there is little academic research which tests whether such interventions have been effective. Together with my Computer Scientist colleague Ranu Malla, we carried out some pilot research to investigate whether *The Guardian*'s visual coverage of heatwaves had changed since their 2019 visual strategy change (see Shields 2019). We examined heatwave coverage as a subset of climate news coverage for two reasons: first, heatwaves are specifically mentioned in *The Guardian*'s visual strategy as a site of problematic visual climate imagery:

> the science tells us a much more sinister story of regular heatwaves and unseasonal weather being a defining indicator of the climate crisis. So, although scenes of children playing in fountains and everyone racing to the beach can be uplifting and irresistible, we have to be mindful of the tone of our journalism. (Shields 2019: np)

And second, we already had a robust coding methodology and comparison dataset from our earlier heatwave visuals work (O'Neill et al 2022).

We followed the methodology in our previous paper (O'Neill et al 2022), which had a database of news images from the northern hemisphere summer 2019, to assemble an equivalent news image dataset from *The Guardian*'s coverage in summer 2022. The results are intriguing. We found that 'fun in the sun' visuals were far rarer in the 2022 dataset; dropping from 8.8 per cent of *The Guardian*'s heatwave coverage in 2019 to 1.4 per cent of similar coverage in 2022. In contrast though, the 'idea of heat' visual frame increased from 8.1 per cent to 19.9 per cent by 2022. Of course, there were notable incidents that contributed to this: the wildfires that broke out in late July 2022 around London, which destroyed some homes, was a particularly newsworthy event (as an aside, it was notable how much more newsworthy the wildfires were, especially in terms of compelling visuals, than the heatwave which preceded it).

We also examined the 2022 dataset for images that portrayed the risk of heat or ways to adapt (see O'Neill et al 2022). We were interested in seeing whether the images depict real people, something specifically called on for in *The Guardian*'s visual strategy (see Shields 2019). The results on this front were underwhelming. In 2022, there was still a substantial lack of images depicting people in the UK at risk of extreme heat. There was not a single image of an older person, for example, and only one showing the adverse impacts of extreme heat on health (depicted visually by an empty hospital bed). Similarly, images did not depict potential solutions or ways to adapt to extreme heat: examples might include images of ways of escaping extreme heat through community cool shelters for at-risk people or through urban greening projects. In conclusion, while *The Guardian*'s visual discourse of heatwaves appeared to be shifting in more obvious ways, there was still be some way to go in depicting the reality of vulnerable people at risk, or in providing a sense of hope and action for adapting to the increasing risk of heat extremes through heatwave visuals.

This is only initial work, but it is worth hypothesising on why this is the case. *The Guardian* is a leading newsroom in terms of thinking through its newsroom strategy for visual climate communication. Yet, even here, there is still some way to go in terms of the work needed to ensure the visual climate discourse is more equitable and representative. One issue may be around scalability and embedded routines of the newsroom. Training for individual journalists and editors assumes that there is some percolation, or spread, of approaches via people's networks. There may be issues around scalability of such approaches to training and education; both within a news organisation, and between that organisation and others. Perhaps more substantially, though, is considering the role of information architecture.

One hypothesis could be questioning the role and power of image agencies in shaping the visual discourse and thinking through the information architectures that bring certain types of image to the fore. While *Guardian* journalists and editors may be searching for different images within *The Guardian* Grid (image management system; Cevey 2015, Hayes and O'Neill 2024) consistent with their visual strategy, how effective are those searches and do the images they search for even exist within the image media ecosystem?

There are opportunities for increasing the diversity of visuals through the creative brief given to photographers, for example; through suggestions for alternative locations or visual stories to capture, as well as in being better prepared for capturing alternative images when climate stories rise up the news agenda. This could be achieved by planning in advance for seasonal events like heatwaves or wildfires, much as newsrooms might plan for political events like the COPs, rather than encountering them as episodic events requiring a short-turnaround response each time. There are also opportunities for thinking through the information architecture of climate imagery; for example, considering how images are tagged, and whether images in a visual database might be tagged differently or additionally to expand the visual discourse of climate change. Journalist Katherine Dunn, content editor at the OCJN, suggested a specific way to increase climate images through tagging: by tagging images of an extreme weather event as climate images, if an attribution study has confirmed that climate change statistically significantly influenced the event's intensity and likelihood (for more information on attribution science, see World Weather Attribution 2024). Such ideas for interventions are worth exploring and testing.

Climate visuals and the growth of generative-AI

This book has sourced and reflected on a diversity of visuals used in climate communications to date. Looking forward, though, brings new challenges, notably around the growth of generative-AI (Gen-AI) tools, which is worth briefly reflecting on. Gen-AI may impact climate visuals in many ways, including through impacts on newsmaking and journalism (Reuters Institute 2024); as well as on the technical, legal, ethical and moral dimensions of climate image-making and engagement (Johnstone 2024).

Gen-AI is showing the capacity to assist and even automate visual news production within journalism (which may result in substantial efficiencies). In terms of how people engage with AI-generated images, they seem able to trigger similar emotional reactions as human-produced images (Paik et al 2023). Gen-AI may offer exciting opportunities for climate imagery, including opportunities to reimagine the visual climate discourse (Sánchez Querubín and Niederer 2022, Johnstone 2024). But substantial challenges exist. Gen-AI is based on large datasets of images and their captions, where systems are trained using machine learning algorithms. Although AI image generators can generate increasingly realistic, aesthetically compelling and high-quality visuals based on simple text prompts (Paik et al 2023), the outputs of an AI image generator are only as good as the real-world datasets of images and captions which are used to train the model in the first place. Gen-AI imagery then risks reinforcing cliched, stereotyped, biased, inaccurate, misleading or unethical visual content (see also Thomas and Thomson 2023). Future research should critically examine emerging tools and practices around Gen-AI and climate visuals, then, especially with regard to a context of financial precarity in the media industry (a precarity which is especially pronounced for the environment beat; Lester 2010).

Towards a more inclusive and responsible climate visual discourse

Twenty years ago, I set off for the University of East Anglia, UEA, to start my PhD. Over 11,000 km away, at the Mauna Loa Observatory in Hawaii, NOAA measured the global average atmospheric carbon dioxide at 377ppm. My PhD was advertised as a project focussing on elucidating the vulnerability and sensitivity upon a small number of global 'icons'. The project was conceived as a new approach for exploring the ultimate objective of international climate policy, as expressed in the UN Framework Convention on Climate Change, to avoid 'dangerous climate change' (see Dessai et al 2004). It was assumed that I would mostly be developing skills in climate modelling; something that, as a Physical Geography undergraduate, I was looking forward to. But as my PhD tenure went on, so the topic focus changed; alongside the literature I consulted and the skills I developed. It became clear that the focus should be on how and why people engaged (or did not) with climate change; rather than the intricacies of modelling climate impacts. Key to the thesis was this idea of 'icons' of climate change: relatable entities people cared about, imagined in their minds, and saw depicted in visual representations (O'Neill and Hulme 2009).

Twenty years on from my PhD, at the Mauna Loa Observatory, atmospheric carbon dioxide has crept ever upwards, measuring 426 ppm as I write this in 2024. While there have been substantial achievements on climate action during those two decades, action is insufficient to address the challenge (UNFCCC 2023). In my own work, I have carried out research with colleagues across diverse disciplines to document and explore how images play a fundamental role in how we imagine transitions and change, such as that needed to address climate change.

This book has discussed and summarised much of that work to show how climate change has something of an image problem. The current climate visual discourse often displaces and marginalises the vulnerability of ecosystems, biodiversity and human communities to climate change. Climate images have also tended to exclude opportunities for imagining a more equitable and resilient future. And, climate images have often been problematic in terms of equity and justice; perpetuating intersecting and ongoing injustices including around gender, race, class, age, poverty and legacies of colonialism. If we are to hope to address the challenge of climate change, we first need to be able to imagine how a different future may look. Images – whether that be photographs, drawings, paintings, films, scientific figures, cartoons and memes or any other type of visual rendering – shape and influence the futures we can imagine, desire and work towards.

The motivation for this book in general, and this concluding chapter in particular, has been to elucidate the work climate images do in the world through an exploration of how images exist and flow in a complex globalised media ecosystem; and how power is exerted through that ecosystem. I have shown how actors, but especially news organisations and image agencies, as well as our own individual preconceptions of what constitutes a climate image, can be powerful in shaping visual representations of climate change. Second, I have shown how power to shape the visual climate discourse is exerted through ways of organising information. Recognising these two interlinked foci also brings to light how there are ways to foreground the experiences of people and places in ways which challenge dominant visual narratives. It is not self-evident that climate change can only be visualised through current, narrow visual tropes: there are ways in which to work towards a more inclusive and responsible climate visual discourse, and which open up other ways to imagine our climate futures.

I increasingly work with and learn from experts outside of academia: many from the media sector such as journalists, editors, filmmakers and photographers; but also people working within organisations such as the IPCC, at different levels in government, at nature agencies, at non-profits and within NGOs and charities. This book is partly testament to those people in sharing their time and expertise over the last couple of decades; but also provides hope that there is a real willingness to recognise and address the visual climate communication challenges I outline here. Thus, this book ends with a call to start to notice (and critique) the climate images that surround us; and to reinvigorate the visuals we use to imagine climate change.

References

350 Pacific (2023) 350 Pacific website. 350.org. https://350.org/pacific/ [accessed 23 June 2023].

Abrahamson, A., Wolf, J., Lorenzoni, I., Fenn, B., Kovats, S., Wilkinson, P., Adger, W.N. and Raine, R. (2009) Perceptions of heatwave risks to health: Interview-based study of older people in London and Norwich, UK. *Journal of Public Health,* 31, 119–126.

Adger, W.N., Lorenzoni, I. and O'Brien, K.L. (2009) *Adapting to climate change: Thresholds, values, governance.* Cambridge University Press.

Adobe (2014) *The Social Intelligence Report – Q1, 2014,* Adobe Digital Index. https://images-template-net.webpkgcache.com/doc/-/s/images.template.net/wp-content/uploads/2016/03/03053721/Social-Media-Intelligence-Report.pdf [accessed 17 April 2024].

Adolphsen, M. and Lück, J. (2012) *Non-routine Interactions behind the scenes of a global media event: How journalists and political PR professionals co-produced the 2010 UN Climate Conference in Cancun.* Medien & Kommunikationswissenschaft, Sonderband 'Grenzüberschreitende Medienkommunikation', 141–158.

AFP (2020) *How to do a reverse image search. AFP: Fact Check.* 6 August 2020. https://factcheck.afp.com/how-do-reverse-image-search [accessed 15 March 2024].

Aiello, G. (2023) Visual communication has always been political. *Journal of Visual Political Communication,* 10(1), 7–16.

Aiello, G. and Parry, K. (2020) *Visual communication.* Sage.

Aiello, G. and Van Leeuwen, T. (2023) Michel Pastoureau and the history of visual communication. *Visual Communication,* 22(1), 27–45.

Aiello, G., Kennedy, H., Anderson, C.W. and Mørk Røstvik, C. (2022) 'Generic visuals' of Covid-19 in the news: Invoking banal belonging through symbolic reiteration. *International Journal of Cultural Studies,* 25(3–4), 309–330.

Ali, A. and Mahmood, S. (2013) Photojournalism and disaster: Case study of visual coverage of Flood 2010 in national newspapers. *Academic Journal of Interdisciplinary Studies,* 2 (9), 168–176. https://doi.org/10.5901/ajis.2013.v2n9p168

Ali, Z.S. (2014) Visual representation of gender in flood coverage of Pakistani print media. *Weather and Climate Extremes*, 4, 35–49. https://doi.org/10.1016/j.wace.2014.04.001

AMOS (2023) AMOS Awards 2023, 21 November 2023. https://www.amos.org.au/amos-awards-2023-2/ [accessed 26 January 2024].

Anderson, C.W. (2013) *Rebuilding the news: Metropolitan journalism in the digital age*. Temple University Press.

Andre, P., Boneva, T., Chopra, F. and Falk, A. (2024) Globally representative evidence on the actual and perceived support for climate action. *Nature Climate Change* 14, 253–259. https://doi.org/10.1038/s41558-024-01925-3

Armstrong, C.L., Boyle, M.P. (2011) *Views from the margins: News coverage of Women Artists for Climate (2024) The Climate Collection*. TED Countdown and Fine Acts. https://artistsforclimate.org/climatecollection [accessed 27 May 2024].

Asian Art Association (2023) South Asia press photo of the Year 2023. Zakir Hossain Chowdhury, Bangladeshi, Anadolu Agency: 'The mother who saved her child in the rainstorm'. https://en.artassociation.asia/app/results/southasia/ [accessed 3 March 2024].

Askanius, T. (2021) Memes and media's role in radicalization. *The Journal of Intelligence, Conflict, and Warfare*, 4(2), 115–121.

Atkins, P. (2022) How Cornish households can protect themselves against rising energy bills. Cornwall Live. 1 April 2022. https://www.cornwalllive.com/special-features/how-cornish-households-can-protect-6889196 [accessed 20 October 2023].

Aubert, M., Brumm, A., Ramli, M., Sutikna, T., Saptomo, E.W., Hakim, B., Morwood, M.J., van den Bergh, G.D., Kinsley, L. and Dosseto, A. (2014) Pleistocene cave art from Sulawesi, Indonesia. *Nature* 514, 223–227.

Azad, M.J. and Pritchard, B. (2023) The importance of women's roles in adaptive capacity and resilience to flooding in rural Bangladesh. *International Journal of Disaster Risk Reduction*, 90, 103660.

Background Stories (2024) Create your own climate generation visual. *Blog*, 8 February. https://backgroundstories.com/create-your-own-climate-generation-visual/ [accessed 3 May 2024].

Badshah, N. (2023) Sadiq Khan hits back at criticism of London Ulez expansion. *The Guardian*. https://www.theguardian.com/environment/2023/sep/03/sadiq-khan-hits-back-at-criticism-of-london-ulez-expansion [accessed 20 October 2023].

Badullovich, N., Grant, W.J. and Colvin, R.M. (2020) Framing climate change for effective communication: A systematic map. *Environmental Research Letters*, 15, 123002.

Banks, M. (2001) *Visual methods in social research*, Sage.

Barendregt, L. (2019) Professor on heat waves and mortality: 'Climate is a threat to health care' [translated from Dutch]. https://www.ad.nl/nijmegen/hoogleraar-over-hittegolven-en-sterfte-klimaat-is-bedreiging-voor-gezondheidszorg~a9e8fc27/ [accessed 26 June 2025].

Barnett, J. and O'Neill, S. (2012) Islands, resettlement and adaptation. *Nature Climate Change*, 2, 8–10.

Barthes, R. (1977) *Image–music–text* [translated by Stephen Heath]. New York.

Batel, S. and Devine-Wright, P. (2021) Using a critical approach to unpack the visual-spatial impacts of energy infrastructures. In S. Batel and D. Rudolph (Eds), *A critical approach to the social acceptance of renewable energy infrastructures – Going beyond green growth and sustainability*. Palgrave Macmillan, 43–60.

BBC (2010) 'Climategate' scientist contemplated suicide, 7 February 2010. http://news.bbc.co.uk/1/hi/uk/8502823.stm [accessed 15 March 2024].

BBC (2021) The trick. https://www.bbc.co.uk/programmes/m0010s10, first aired 18 October 2021 [accessed 15 March 2024].

BBC (2022) UK heatwave: Temperatures to hit low 30s as heat-health alert issued. https://twitter.com/SaffronJONeill/status/1545290526336884738 [accessed 8 July 2022].

BBC (2023) What do colours on the BBC Weather maps mean? *BBC Weather*, 25 July. https://www.bbc.co.uk/weather/features/66293839 [accessed 24 November 2023].

Beck, U. (1992) *Risk society: Towards a new modernity*. Sage.

Bednarek, M. and Caple, H. (2012) 'Value added': Language, image and news values. *Discourse, Context, Media*, 1: 103–113.

Behrmann, A. (2019) The artists of Extinction Rebellion: 'Our bold imagery is helping to change the conversations around climate change'. iNews, 24 November 2019. https://inews.co.uk/culture/arts/extinction-rebellion-artist-protest-banner-art-red-rebel-flag-logo-366404 [accessed 2 January 2024].

BEIS (2021) Net zero strategy: Build back greener. Presented to Parliament pursuant to Section 14 of the Climate Change Act 2008. https://assets.publishing.service.gov.uk/media/6194dfa4d3bf7f0555071b1b/net-zero-strategy-beis.pdf [accessed 30 June 2025].

Benjamin, P. (2019) The meaning behind Extinction Rebellion's red-robed protesters. Dazed, 26 April. https://www.dazeddigital.com/politics/article/44238/1/meaning-behind-extinction-rebellions-red-robed-protesters-london-climate-change [accessed 3 January 2024].

Bennet, C., Foley, M. and Krebs, H.B. (2016) Learning from the past to shape the future: Lessons from the history of humanitarian action in Africa. *Humanitarian Policy Group Working Paper*. Overseas Development Institute.

Bennett, W.L. (2011) *News: The politics of illusion*. Longman, 9th ed.

Berger, J. (1972) *Ways of seeing*. BBC and Penguin Books.

Birnbaum, M. and Kaplan, S. (2022) Climate deal is a 'baby step,' but world needs bigger action, diplomats say. Washington Post, 29 July 2022. https://www.washingtonpost.com/climate-environment/2022/07/29/international-activists-diplomats-react-climate-bill/ [accessed 23 June 2023].

Bissell, K. (2000a) Culture and gender as factors in photojournalism gatekeeping. *Visual Communication Quarterly*, 7(2), 9–12.

Bissell, K.L. (2000b) A return to 'Mr. Gates': Photography and objectivity. *Newspaper Research Journal*, 21(3), 81–93.

Bleiker, R. (2018) Mapping visual global politics. In R. Bleiker (Ed), *Visual global politics.* Routledge.

BOM (2022) Special Climate Statement 76 – Extreme rainfall and flooding in south-eastern Queensland and eastern New South Wales. *Australian Government Bureau of Meteorology*, 25 May 2022. http://www.bom.gov.au/climate/current/statements/scs76.pdf?20220525 [accessed 26 July 2023].

Born, D. (2018) Bearing witness? Polar bears as icons for climate change communication in National Geographic. *Environmental Communication*, 13, 649–663. https://doi.org/10.1080/17524032.2018.1435557

Boussalis, C., Coan, T.G., Holman, M.R. and Müller, S. (2021) Gender, candidate emotional expression, and voter reactions during televised debates. *American Political Science Review* 115(4), 1242–1257.

Boussalis, C., Coan, T.G., Malla, R. and O'Neill, S. (2024) *Diverging perspectives? Comparing visual heatwave portrayals by developed and developing country media outlets.* 2024 COMPTEXT Conference, 2–4 May, Vrije Universiteit.

Boyle, L. (2023) Kathy Lette forced to apologise after 'unacceptable' coronation joke about sinking Pacific island of Tuvalu. *The Independent* 8 May 2023. https://www.independent.co.uk/climate-change/news/royal-coronation-kathy-lette-tuvalu-b2334924.html [accessed 13 October 2023].

Briggs, L. (2003) Mother, child, race, nation: The visual iconography of rescue and the politics of transnational and transracial adoption. *Gender & History*, 15, 179–200.

Brimicombe, C., Porter, J.J., Di Napoli, C., Pappenberger, F., Cornforth, R., Petty, C. and Cloke, H.L. (2021) Heatwaves: An invisible risk in UK policy and research. *Environmental Science & Policy*, 116, 1–7.

Broz, M. (2023) *How many pictures are there (2024): Statistics, trends, and forecasts*, 24 November 2023. https://phototutorial.com/photos-statistics/ [accessed 9 January 2024].

Buckland, E. (2023) Before the floods I thought climate change wasn't my problem. Now, I'm not waiting for someone else to fix it. *The Guardian*, 28 February 2023. https://www.theguardian.com/commentisfree/2023/feb/28/before-the-floods-i-thought-climate-change-wasnt-my-problem-now-im-not-waiting-for-someone-else-to-fix-it [accessed 26 July 2023].

Bulkeley, H., Paterson, M. and Stripple, J. (Eds) (2016) *Towards a cultural politics of climate change: Devices, desires and dissent.* Cambridge University Press.

Burgess, K. (2021) IPCC climate change report: Summary of a wake-up call for the world. *The Times,* 9 August 2023. https://www.thetimes.co.uk/article/ipcc-climate-change-report-a-wake-up-call-for-the-world-l73clz7nf#:~:text=The%20sea%20level%20will%20rise,the%20end%20of%20the%20century [accessed 17 November 2023].

Bush, G.W. (2008). President Bush discusses climate change, Rose Garden, 2:45 P.M. EDT.

Campbell, D. (2010) Geopolitics and visuality: Sighting the Darfur conflict. *Political Geography*, 26, 357–382. https://doi.org/10.1016/j.polgeo.2006.11.005

Caple, H. (2013) *Photojournalism: A social semiotic approach.* Palgrave Macmillan.

Caple, H. and Bednarek, M. (2016) Rethinking news values: What a discursive approach can tell us about the construction of news discourse and news photography. *Journalism*, 17(4), 435–455.

Capstick, S., Whitmarsh, L., Poortinga, W., Pidgeon, N. and Upham, P. (2015) International trends in public perceptions of climate change over the past quarter century. *WIREs Climate Change*, 6, 35–61. https://doi.org/10.1002/wcc.321

Carbon Brief (2022) Media reaction to 40°C record-breaking heat via newspaper frontpages image collage. https://www.carbonbrief.org/wp-content/uploads/2022/07/hero-montage-1550x804.png [accessed 13 October 2023].

Carbon Brief (2024) The Global South Climate Database. https://www.carbonbrief.org/global-south-climate-database/ [accessed 10 January 2024].

Carvalho, A. and Burgess, J. (2005) Cultural circuits of climate change in UK broadsheet newspapers, 1985–2003. *Risk Analysis*, 25, 1457–1469.

Catanzaro, M. and Collin, P. (2023) Kids communicating climate change: Learning from the visual language of the SchoolStrike4Climate protests. *Educational Review*, 75(1), 9–32.

Cave, S. and Dihal, K. (2020) The whiteness of AI. *Philosophy and Technology*, 33, 685–703.

CBD (2018, 24 April) *Saving the polar bear.* Centre for Biological Diversity. https://www.biologicaldiversity.org/species/mammals/polar_bear/ [accessed 3 July 2025].

Cevey, S. (2015, 12 August) Open sourcing grid, the Guardian's new image management service. *The Guardian.* https://www.theguardian.com/info/developer-blog/2015/aug/12/open-sourcing-grid-image-service [accessed 3 July 2025].

Chadwick, A. (2017) *The Hybrid Media System: Politics and Power,* 2nd ed. Oxford University Press.

Chambers, D.W. (1983) Stereotypic images of the scientist: The A-Draw-A-Scientist Test. *Science Education*, 67(2), 255–265.

Chan, J. and Lee, C. (1984) Journalistic paradigms on civil protests: A case study of Hong Kong. In A. Arno and W. Dissanayake (Eds), *The News Media in National and International Conflict*. Westview Press, 183–202.

Chapman, D.A., Corner, A., Webster, R. and Markowitz, E.M. (2016) Climate visuals: A mixed methods investigation of public perceptions of climate images in three countries. *Global Environmental Change*, 41, 172–182.

Chowdhury, A.M.R., Bhuyia, A.U., Choudhury, A.Y. and Sen, R. (1993) The Bangladesh cyclone of 1991: why so many people died. *Disasters*, 17(4), 291–304.

Climate Central (2024) Picturing our future: Climate and energy choices this decade will influence how high sea levels rise for hundreds of years. Which future will we choose? https://picturing.climatecentral.org/ [accessed 5 July 2024].

Climate Outreach (2024) Climate visuals: A Climate Outreach project (website). https://www.climatevisuals.org/climate-visuals/ [accessed 30 June 2025].

ClimateShitPost (2023) My take on the IPCC chart. Reddit, r/ClimateShitposting. https://www.reddit.com/r/ClimateShitposting/comments/11xhnlq/my_take_on_the_ipcc_chart/ [accessed 9 May 2024].

Coan, T.G., Boussalis, C., Cook, J. and Nanko, M. (2021) Computer-assisted classification of contrarian claims about climate change. *Nature Scientific Reports*, 11, 22320.

Coca-Cola (2021, 16 March). Coca-Cola's polar bears: An enduring legacy. https://www.coca-colacompany.com/company/history/coca-colas-polar-bears [accessed 3 July 2025].

Coleman, R. (2010) Framing the pictures in our heads. In P. D'Angelo and J.A. Kuypers (Eds), *Doing news framing analysis: Empirical and theoretical perspectives*. Routledge, 233–261.

Colose, C. (2023) I think an actual unpopular opinion – I really like the direction IPCC wants to go in with visual elegance in its presentation, but … *21 March*. https://twitter.com/CColose/status/1638020392039186434 [accessed 24 November 2023].

Corner, A. (2022) This is what the transition looks like: Introducing the local storytelling exchange. Medium. https://medium.com/inter-narratives/this-is-what-the-transition-looks-like-introducing-the-local-storytelling-exchange-eb0d053bcc4d [accessed 20 October 2023].

Corner, A., Webster, R. and Teriete, C. (2015). *Climate visuals: Seven principles for visual climate change communication (based on international social research)*. Climate Outreach.

REFERENCES

Cosgrove, D. (1994) Contested global visions: One-world, whole-Earth and the Apollo space photographs. *Annals of the American Association of Geographers,* 84, 270–294.

Cosgrove, D. and della Dora V. (2008). *High places: Cultural geographies of mountains and ice.* Tauris Books.

Craymer, L. (2022) Digital replica: Tuvalu turns to Metaverse to guarantee its existence. *Sydney Morning Herald,* 17 November 2022. https://www.smh.com.au/world/oceania/digital-replica-tuvalu-turns-to-metaverse-to-guarantee-its-existence-20221117-p5bz4c.html [accessed 23 June 2023].

Crockford, S. (2015) Twenty good reasons not to worry about polar bears. *The Global Warming Policy Foundation.* https://www.thegwpf.org/content/uploads/2015/02/Crockford-2015.pdf [accessed 26 July 2023].

Cronon, W. (1996) *Uncommon ground: Rethinking the human place in nature.* W.W. Norton and Company, 69–90.

Covering Climate Now (2023) Climate Solutions Reporting Guide, 9 January 2023. https://coveringclimatenow.org/resource/climate-solutions-reporting-guide/ [accessed 5 July 2024].

Dancygier, B. and Vandelanotte, L. (2017) Internet memes as multimodal constructions. *Cognitive Linguistics,* 28(3), 565–598.

Davis, C.B., Glantz, M. and Novak, D.R. (2016) 'You can't run your SUV on cute. Let's go!': Internet memes as delegitimizing discourse. *Environmental Communication,* 10(1), 62–68.

Davoudi, S. and Machen, R. (2022) Climate imaginaries and the mattering of the medium. *Geoforum,* 137, 203–212.

De Smaele, H., Geenen, E. and De Cock, R. (2017) Visual gatekeeping – Selection of news photographs at a Flemish newspaper: A qualitative inquiry into the photo news desk. *Nordicom Review,* 38(2), 57–70.

Derocher, A.E., Lunn, N.J. and Stirling, I. (2004) Polar bears in a warming climate. *Integrative and Comparative Biology,* 44, 163–176.

Dessai, S., Adger, W.N., Hulme, M., Turnpenny, J., Köhler, J. and Warren, R. (2004) Defining and experiencing dangerous climate change. *Climatic Change,* 64, 11–25.

Devine-Wright, H. (2014) Envisioning public engagement with renewable energy: An empirical analysis of images within the UK national press 2006/2007. In P. Devine-Wright (Ed) *Renewable Energy and the Public,* Routledge, 101–113.

Devine-Wright, H. and Devine-Wright, P. (2009) Social representations of electricity network technologies: Exploring processes of anchoring and objectification through the use of visual research methods. *British Journal of Social Psychology,* 48 (2), 357–373.

Devine-Wright, P., Whitmarsh, L., Gatersleben, B., O'Neill, S., Hartley, S., Burningham, K., Sovacool, B., Barr, S. and Anable, J. (2022) Placing people at the heart of climate action. *PLOS Climate,* 1(5), e0000035.

Devine-Wright, P. and Wiersma, B. (2021) Auto-photography, senses of place and public support for marine renewable energy. In C.M. Raymond, L.C. Manzo, D.R. Williams, A.D. Masso, A.D. and T. von Wirth (Eds), *Changing senses of place: Navigating global challenges*. Cambridge University Press. 144–155.

Diaz, V.M. and Kauanui, J.K. (2001) Native Pacific cultural studies on the edge. *The Contemporary Pacific*, 13.2, 315–342.

DiFrancesco, D.A. and Young, N. (2011) Seeing climate change: The visual construction of global warming in Canadian national print media. *Cultural Geographies*, 18(4), 517–536.

Digital Desk Staff (2021) Support grows among voters for climate change action, poll shows. *Breaking News Ireland*, 13 December 2021. https://www.breakingnews.ie/ireland/support-grows-among-voters-for-climate-change-action-poll-shows-1227449.html [accessed 23 June 2023].

Dihal, K. and Duarte, T. (2023) *Better images of AI: A guide for users and creators*. The Leverhulme Centre for the Future of Intelligence and We and AI.

Dixon, D. (2023) In the breach: Feeling the heat of climate change, *Scottish Geographical Journal*, 139(1–2): 103–114.

Dodds, T., de Vreese, C., Helberger, N., Resendez, V. and Seipp, T. (2023) Popularity-driven metrics: Audience analytics and shifting opinion power to digital platforms. *Journalism Studies*, 24(3).

Doherty, B. (2020) Australia should create 'Pacific visa' to reduce impact of CC and disaster on islanders. *The Guardian*, 21 October 2020. https://www.theguardian.com/world/2020/oct/21/australia-should-create-pacific-visa-to-reduce-impact-of-climate-change-and-disaster-on-islanders#:~:text=Australia%20should%20establish%20a%20new,Australia's%20region%20in%20coming%20decades [accessed 23 June 2023].

Domke, D., Perlmutter, D. and Spratt, M. (2002) The prime of our times? An examination of the 'power' of visual images. *Journalism*, 3(2), 131–159.

Donner, S. (2020) The ugly history of climate determinism is still evident today. *Scientific American*. 24 June 2020. https://www.scientificamerican.com/article/the-ugly-history-of-climate-determinism-is-still-evident-today/ [accessed 27 May 2024].

Doyle, J. (2007) Picturing the clima(c)tic: Greenpeace and the representational politics of climate change communication. *Science as Culture*, 16, 129–150.

Doyle, J. (2009) Seeing the climate? The problematic status of visual evidence in climate change campaigning. In S. Dobrin and S. Money (Eds), *Ecossee: Images, rhetoric and nature*. State University of New York Press, 279–297.

Doyle, J. (2011) *Mediating climate change*. Routledge.

Duan, R., Takahashi, B. and Zwickle, A. (2019) Abstract or concrete? The effect of climate change images on people's estimation of egocentric psychological distance. *Public Understanding of Science*, 28(7), 828–844.

Dunaway, F. (2015) *Seeing green: The use and abuse of American environmental images*. University of Chicago Press.

Dyer, G. (1982) *Advertising as communication*. Routledge.

Earth Island Institute (2022) As Europe suffers 'heat apocalypse', UK smashes temperature record. https://www.earthisland.org/journal/index.php/articles/entry/europe-suffers-heat-apocalypse-uk smashes-temperature-record/ [accessed 23 June 2023].

Edwards, C. (2022) The role and the risks of solutions journalism for climate reporting. *10 August 2022*. https://www.journalism.co.uk/news/the-role-and-the-risks-of-solutions-journalism-for-climate-reporting/s2/a954292/ [accessed 5 July 2024].

Eide, E. (2012) Visualizing a global crisis. Constructing climate, future and present. *Conflict & Communication Online*, 11.

Ejaz, W., Mukherjee, M. and Fletcher, R. (2023) Climate change news audiences: Analysis of news use and attitudes in eight countries. *Oxford Climate Journalism Network, Reuters Institute for the Study of Journalism*, University of Oxford.

Ejaz, W. and Najam, A. (2023). The Global South and climate coverage: From news taker to news maker. *Social Media + Society*, 9(2).

Engelhard, M. (2016). *Ice bear: The cultural history of an Arctic icon*. University of Washington Press.

Engesser, S. and Brüggemann, M. (2015) Mapping the minds of the mediators: The cognitive frames of climate journalists from five countries. *Public Understanding of Science*, 25, 825–841.

Entman, R.M. (1993) Framing: Toward clarification of a fractured paradigm. *Journal of Communication*, 43, 51–58.

Entman, R.M., Matthes, J. and Pellicano, L. (2009) Nature, sources, and effects of news framing. In K. Wahl-Jorgensen and T. Hanitzsch (Eds), *The Handbook of Journalism Studies*. Routledge, 175–190.

Erscoi, L.A., Kleinherenbrink, A. and Guest, O. (2023) Pygmalion displacement: When humanising AI dehumanises women. *SocArXiv*. doi: 10.31235/osf.io/jqxb6.

European Parliament (2019) Women, gender equality and the energy transition in the EU. Policy Department for Citizens' Rights and Constitutional Affairs, Directorate General for Internal Policies of the Union, *PE* 608.867 (May 2019). https://www.europarl.europa.eu/RegData/etudes/STUD/2019/608867/IPOL_STU(2019)608867_EN.pdf [accessed 5 January 2024].

Evans. J. and Hall, S. (1999) *Visual Culture: The Reader*, 2nd ed. Sage, in association with the Open University.

Evelyn, K. (29/01/2020) 'Like I wasn't there': Climate activist Vanessa Nakate on being erased from a movement. *The Guardian*, 11 September 2021. www.theguardian.com/world/2020/jan/29/vanessa-nakate-interview-climate-activism-cropped-photo-davos.

Fahmy, S. (2005) Photojournalists' and photo editors' attitudes and perceptions: The visual coverage of 9/11 and the Afghan war. *Visual Communication Quarterly,* 12(3–4), 146–163.

Fahmy, S., Bock, M. and Wanta, W. (2014) *Visual communication theory and research: A mass communication perspective*. Palgrave Macmillan.

Fahmy, S., Kelly, J.D. and Kim, Y.S (2007) What Katrina revealed: A visual analysis of the hurricane coverage by news wires and US newspapers. *Journalism & Mass Communication Quarterly*, 84(3), 546–561.

Farbotko, C. (2010) Wishful sinking: Disappearing islands, climate refugees and cosmopolitan experimentation. *Asia Pacific Viewpoint,* 51, 47–60.

Farbotko, C., Boas, I., Dahm, R., Kitara, T., Lusama, T. and Tanielu, T. (2023) Reclaiming open climate adaptation futures. *Nature Climate Change*, 13, 750–751.

Farbotko, C. and Campbell, J. (2022) Who defines atoll 'uninhabitability'? *Environmental Science & Policy*, 138, 182–190.

Farbotko, C. and Kitara, T. (2022) Climate leadership in the 'disappearing islands'. *Georgetown Journal of International Affairs.* https://gjia.georgetown.edu/2022/05/06/climate-leadership-in-the-disappearing-islands%EF%BF%BC/ [accessed 30 June 2025].

Figueroa, E.J. (2022). Casting heroes and victims of disaster events: Representations of race and gender in Hurricane Harvey front page news images. *Critical Studies in Media Communication*, 39(5), 455–471. https://doi.org/10.1080/15295036.2022.2121412

Fischer, H., Schütte, S., Depoux, A., Amelung, D. and Sauerborn, R. (2018) How well do COP22 attendees understand graphs on climate change health impacts from the fifth IPCC assessment report? *International Journal of Environmental Research and Public Health,* 15(5), 875.

Fischer, H., van den Broek, K., Ramisch, K. and Okan, Y. (2020) When IPCC graphs can foster or bias understanding: Evidence among decision-makers from governmental and non-governmental institutions. *Environmental Research Letters,* 15, 114041.

Fletcher, R., Eddy, K., Robertson, C.T. and Nielsen, R.K. (2023) Tuvalu Coastal Adaptation Project album (243 photos). *Flickr.* https://www.flickr.com/photos/undpclimatechangeadaptation/albums/72157687817446872/ [accessed 11 December 2023].

Flockemann, R. (2023) Comic books aren't just about superheroes: They can also be a great way of communicating research. Sussex Energy Group at SPRU blog post. 31 July 2023. https://blogs.sussex.ac.uk/sussexenergygroup/2023/07/31/comic-books-arent-just-about-superheros-they-can-also-be-a-great-way-of-communicating-research/ [accessed 10 November 2023].

Fox, K. (2005) *Watching the English: Hidden rules of English behaviour.* Hodder Stoughton.

Francisco, D. (2023) *Red Rebel Brigade.* https://dougfrancisco.com/red-rebel-brigade/ [accessed 02 January 2024].

Freelon, D. (2013) Discourse architecture, ideology, and democratic norms in online political discussion. *New Media & Society*, 17(5), 772–791. https://doi.org/10.1177/1461444813513259

Frosh, P. (2003) *The image factory: Consumer culture, photography and the visual content industry.* Berg.

Frosh P. (2020) Is commercial photography a public evil? Beyond the critique of stock photography. In M. Miles and E. Welch (Eds), *Photography and its publics.* Bloomsbury, pp 187–206.

Fruean, B. (2021) Pacific islanders aren't just victims – we know how to fight the climate crisis. *The Guardian*, 2 November 2021. https://www.theguardian.com/commentisfree/2021/nov/02/pacific-islanders-fight-climate-crisis-cop26 [accessed 23 June 2023].

Gabbatiss, J. (2018) Climate Change Act must set 'net zero' emissions target by 2020, experts say. *The Independent*, 30 March 2018. https://www.independent.co.uk/climate-change/news/climate-change-act-net-zero-emissions-target-experts-greenhouse-gases-uk-a8280606.html [accessed 17 November 2023].

Galtung, J. and Ruge, M.H. (1965) The structure of foreign news: The presentation of the Congo, Cuba and Cyprus crises in four Norwegian newspapers. *Journal of Peace Research,* 2(1), 64–90. https://doi.org/10.1177/002234336500200104

Garfield, S. (2007) Can polar bears save the world? *The Observer Magazine,* 4 March 2007.

GBNews (2023) Neil Oliver: 'Fear mongering' over high temperatures is an 'incessant attempt to keep us frightened'. 17 July. https://www.youtube.com/watch?v=8vkpcBMmvP0 [accessed 24 November 2023].

Getty Images (2023) 'Family lifestyle travel wind energy' 15-image collection, Me 3645 Studio. https://www.gettyimages.co.uk/search/stack/810685940?family=creative&assettype=image [accessed 17 November 2023].

Getty Images (2024a) Photograph titled: 'Sustainable power is the future', by ljubaphoto, Creative: #1316576079, uploaded 7 May 2021. https://www.gettyimages.co.uk/search/stack/802779172?family=creative&assettype=image [accessed 11 April 2024].

Getty Images (2024b) Getty Images keyword guide. http://www.guliver.ro/wp-content/uploads/Guliver_-_Getty_Images_Keyword_Guide.pdf. but accessible via the WayBack Machine: https://web.archive.org/web/20230202193008/http://www.guliver.ro/wp-content/uploads/Guliver_-_Getty_Images_Keyword_Guide.pdf, captured 2 Feb 2023 [accessed 03 July 2025]

Giannoli, V. (2023) Giornata dell'ambiente, udienza del Papa con il gruppo Gedi: 'La difesa del clima è un bene comune'. *La Repubblica*, 5 June 2023. https://www.repubblica.it/green-and-blue/dossier/festival-greenandblue-2023/2023/06/05/news/papa_francesco_molinari_ambiente_clima-403260554/ [accessed 26 January 2024].

Gleick, P.H. et al (2010). Climate change and the integrity of science. *Science*, 328, 689–690. https://doi.org/10.1126/science.328.5979.689

Goldberg, S. (2016) How we spot altered pictures. *National Geographic*. https://www.nationalgeographic.com/magazine/article/editors-note-images-and-ethics [accessed 26 July 2023].

Gommeh, E., Dijstelbloem, H. and Metze, T. (2021) Visual discourse coalitions: Visualization and discourse formation in controversies over shale gas development, *Journal of Environmental Policy & Planning*. doi: 10.1080/1523908X.2020.1823208.

Goodhew, J., Pahl, S., Auburn, T. and Goodhew, S. (2015) Making heat visible: Promoting energy conservation behaviors through thermal imaging. *Environment and Behavior*, 47(10), 1059–1088.

Grabe, M.E. and Bucy, E.P. (2009) *Image bite politics: News and the visual framing of elections*. Oxford University Press.

Graber, D.A. (1990) Seeing is remembering: How visuals contribute to learning from television news. *Journal of Communication*, 40(3), 134–155.

Grattan, M. (2010) Greens secure position to put their footprint all over carbon decisions. *The Age*, 28 September 2010. https://www.theage.com.au/politics/federal/greens-secure-position-to-put-their-footprint-all-over-carbon-decisions-20100927-15u40.html [accessed 02 January 2024].

Greer, B. (2014). *Craftivism: The art of craft and activism*. Arsenal Pulp Press.

Grittman, E. (2014) Between beauty, risk and the sublime: The visualisation of climate change in media coverage during COP 15 in Copenhagen 2009. In B. Schneider and T. Nocke (Eds), *Image politics of climate change*. Transcript Verlag.

Guggenheim, D. (2006) An inconvenient truth. Lawrence Bender Productions, Participant Productions.

Gynnild, A. (2017) The visual power of news agencies. *Nordicom Review*, 38, 25–39.

Hadland, A. and Barnett, C. (2018) *The state of news photography: Photojournalists' attitudes toward work practices, technology and life in the digital age*. World Press Photo and University of Stirling. https://www.worldpressphoto.org/getmedia/4f811d9d-ebc7-4b0b-a417-f119f6c49a15/the_state_of_news_photography_2018.pdf [accessed 22 March 2024].

Hall, S. (1997) The work of representation. In S. Hall (Ed), *Representation: Cultural representations and signifying practices*. Sage, 13–74.

Hall, S. (1973) The determinations of news photographs. In S. Cohen and J. Young (Eds), *The manufacture of news: Deviance, social problems and the mass media*. Constable, 176–190.

Hall, S. (1980) Encoding/decoding. In S. Hall, D. Hobson, A. Lowe and P. Willis (Eds), *Culture, media, language*. Hutchinson, 128–138.

Hambly, E. (2022) COP27: Climate change threatening global health – report. *BBC News website*, 26 October 2022. https://www.bbc.co.uk/news/science-environment-63386814 [accessed 17 November 2023].

Hansen, A. and Machin, D. (2008) Visually branding the environment: climate change as a marketing opportunity. *Discourse Studies*, 10(6), 777–794.

Hansen, A. and Machin, D. (2013) Researching visual environmental communication. *Environmental Communication*, 7(2), 151–168. https://doi.org/10.1080/17524032.2013.785441

Harcup, T. and O'Neill, D. (2001) What is news? Galtung and Ruge revisited. *Journalism Studies*, 2(2), 261–280.

Harcup, T. and O'Neill, D. (2016) What is news? News values revisited (again). *Journalism Studies*, 18(12), 1470–1488.

Harold, J., Lorenzoni, I., Shipley, T. and Coventry, K.R. (2016) Cognitive and psychological science insights to improve climate change data visualization. *Nature Climate Change*, 6, 1080–1089.

Hart, P.S., Feldman, L., Choi, S., Zhang, A.L. and Hegland, A. (2023) The influence of flooding imagery and party cues on perceived threat, collective efficacy, and intentions for political action to address climate change. *Science Communication*, 45(5), 627–664.

Hart, S. and Feldman, L. (2016) The impact of climate change-related imagery and text on public opinion and behavior change. *Science Communication*, 38, 415–441.

Hase, V., Mahl, D., Schäfer, M.S. and Keller, T.R. (2021) Climate change in news media across the globe: An automated analysis of issue attention and themes in climate change coverage in 10 countries (2006–2018). *Global Environmental Change*, 70, 102353.

Hau'ofa, E. (1994) Our sea of islands. *The Contemporary Pacific*, 6(1), 148–161.

Hautea, S., Parks, P., Takahashi, B. and Zeng, J. (2021). Showing they care (or don't): Affective publics and ambivalent climate activism on TikTok. *Social Media + Society*, 7(2).

Hawkins, E. (2016) *Spiralling global temperatures from 1850–2016 (full animation)*. 9 May. https://twitter.com/ed_hawkins/status/729753441459945474 [accessed 24 November 2023].

Hawkins, E. (2023) *Yesterday the IPCC 6th Assessment published its Synthesis Report (SYR) which included this graphic (SYR.1c)*. 21 March. https://twitter.com/ed_hawkins/status/1638117210303483905 [accessed 24 November 2023].

Hawkins, E., Fæhn, T. and Fuglestvedt, J. (2019) The climate spiral demonstrates the power of sharing creative ideas. *Bulletin of the American Meteorological Society*, 100(5), 753–756.

Hayes, S. (2024) *'How am I supposed to do this?': Navigating the use of visuals in the production of digital climate change journalism.* Unpublished PhD thesis, Geography, University of Exeter.

Hayes, S. and O'Neill, S. (2021) The Greta effect: Visualising climate protest in UK media and the Getty Images collections. *Global Environmental Change*, 71, 102392.

Hayes, S. and O'Neill, S. (2024) Visual politics, protest and power: Who shaped the climate visual discourse at COP26? *Journalism Studies*, 26(4), 441–463.

Heintz, B. and Huber, J. (Eds) (2001) *Mit dem Auge denken. Strategien der Sichtbarmachung in wissenschaftlichen und virtuellen.* Welten Springer.

Highwood, E. (2017a) *OK so I am a crochet addict [Twitter] 10 June.* https://twitter.com/EllieHighwood/status/873603203279015937?lang=en [accessed 24 November 2023].

Highwood, E. (2017b) *#climatechangecrochet – The global warming blanket [blog post].* 12 June. https://elliehighwood.com/2017/06/12/climatechangecrochet-the-global-warming-blanket/ [accessed 24 November 2023].

Hinchcliffe, J. (2022) Adrift in a parallel universe: Lismore struggles to find itself after another flood. *The Guardian*, 1 April 2022. https://www.theguardian.com/australia-news/2022/apr/02/adrift-in-a-parallel-universe-lismore-struggles-to-find-itself-after-another-flood [accessed 26 July 2023].

Hopke, J.E. and Hestres, L.E. (2018) Visualizing the Paris Climate Talks on Twitter: Media and Climate Stakeholder Visual Social Media During COP21. *Social Media + Society*, 4(3).

House of Lords (2022) Corrected oral evidence: Mobilising action on climate change and environment: behaviour change. *Environment and Climate Change Committee*. 27 April 2022. https://committees.parliament.uk/oralevidence/10181/html/ [accessed 20 October 2023].

Huang, Q. (2011) A study on the metaphor of 'red' in Chinese culture, *American International Journal of Contemporary Research*, 1(3), 99–102.

Berger, John. (1972). Ways of Seeing. London: BBC and Penguin Books., M. (2009) *Why we disagree about climate change.* Cambridge University Press.

Huntington, H.P., Carey, M., Apok, C., Forbes, B.C., Fox, S., Holm, L.K., Ivanova, A., Jaypoody, J., Noongwook, G. and Stammler, F. (2019) Climate change in context: putting people first in the Arctic. *Regional Environmental Change*, 19, 1217–1223.

IPCC (2001a) *Climate change 2001: impacts, adaptation and vulnerability, Contribution of Working Group II to the Third Assessment Report of the Intergovernmental Panel on Climate Change.* J.J. McCarthy, O.F. Canziani, N.A. Leary, D.J. Dokken and K.S. White (Eds). Cambridge University Press.

REFERENCES

IPCC (2001b) Climate change 2001: Synthesis report, Contribution of Working Groups I, II, and III to the Third Assessment Report of the Intergovernmental Panel on Climate Change. R.T. Watson and the Core Writing Team (Eds). Cambridge University Press.

IPCC (2014) *Climate Change 2014: Impacts, Adaptation, and Vulnerability. Contribution of Working Group II to the Fifth Assessment Report of the Intergovernmental Panel on Climate Change.* C.B. Field, V.R. Barros, D.J. Dokken, K.J. Mach, M.D. Mastrandrea, T.E. Bilir, M. Chatterjee, K.L. Ebi, Y.O. Estrada, R.C. Genova, B. Girma, E.S. Kissel, A.N. Levy, S. MacCracken, P.R. Mastrandrea and L.L. White (Eds) Cambridge University Press.

IPCC (2021) *Climate Change 2021: The Physical Science Basis. Contribution of Working Group I to the Sixth Assessment Report of the Intergovernmental Panel on Climate Change.* V. Masson-Delmotte, P. Zhai, A. Pirani, S.L. Connors, C. Péan, S. Berger, N. Caud, Y. Chen, L. Goldfarb, M.I. Gomis, M. Huang, K. Leitzell, E. Lonnoy, J.B.R. Matthews, T.K. Maycock, T. Waterfield, O. Yelekçi, R. Yu and B. Zhou (Eds). Cambridge University Press.doi:10.1017/9781009157896.

IPCC (2022) *Climate Change 2022: Impacts, Adaptation and Vulnerability. Contribution of Working Group II to the Sixth Assessment Report of the Intergovernmental Panel on Climate Change.* H-O Pörtner, D.C. Roberts, M. Tignor, E.S. Poloczanska, K. Mintenbeck, A. Alegría, M. Craig, S. Langsdorf, S. Löschke, V. Möller, A. Okem and B. Rama (Eds). Cambridge University Press, 1817–1927.

IPCC (2023) *Climate Change 2023: Synthesis Report. A Report of the Intergovernmental Panel on Climate Change. Contribution of Working Groups I, II and III to the Sixth Assessment Report of the Intergovernmental Panel on Climate Change* H. Lee and J. Romero (Eds). IPCC.

iStock (2023) ChameLeón seye Stock image and video portfolio – iStock. 309 Fiji Pictures, Images and Stock Photos. https://www.istockphoto.com/portfolio/chameLeón seye?assettype=image&mediatype=photography&phrase=fiji&alloweduse=off [accessed 4 July 2023].

IUCN. (2015) New assessment highlights climate change as most serious threat to polar bear survival – IUCN Red List. https://www.iucn.org/content/new-assessment-highlights-climate-change-most-serious-threat-polar-bear-survival-iucn-red [accessed 11 July 2023].

Jaakonmäki, R., Müller, O. and vom Brocke, J. (2017, January). The impact of content, context, and creator on user engagement in social media marketing. In Proceedings of the 50th Hawaii International Conference on System Sciences. https://pdfs.semanticscholar.org/2806/75b30795b13752b8e01d825c29626351d044.pdf [accessed 30 June 2025].

Jarillo, S. and Barnett, J. (2022) Repositioning the (is)land: Climate change adaptation and the atoll assemblage. *Antipode,* 54, 848–872.

Joffe, H. (2008) The power of visual material: Persuasion, emotion and identification. *Diogenes,* 55(1), 84–93.

Johnstone, A. (2024) What does AI imagery mean for climate change photography? Climate visuals, climate outreach. https://climateoutreach.org/ai-climate-change-photography/ [accessed 17 April 2024].

Jones, L. and Boyd, E. (2011) Exploring social barriers to adaptation: Insights from Western Nepal. *Global Environmental Change,* 21(4), 1262–1274.

Kahle, S., Yu, N. and Whiteside, E. (2007). Another disaster: An examination of portrayals of race in Hurricane Katrina coverage. *Visual Communication Quarterly,* 14(2), 75–89. https://doi.org/10.1080/15551390701555951

Kaminski, I. (2023) Children's voices must be heard on climate crisis, says UN rights body. *The Guardian,* 28 August 2023. https://www.theguardian.com/environment/2023/aug/28/childrens-voices-must-be-heard-on-climate-crisis-says-un-rights-body [accessed 15 February 2024].

Karlsson, M. and Clerwall, C. (2013) Negotiating professional news judgment and 'clicks' comparing tabloid, broadsheet and public service traditions in Sweden. *Nordicom Review,* 34(2), 65–76.

Katz, A. (2017, 22 December). Top 10 photos of the year: From the Editors of TIME. TIME Magazine. http://time.com/top-10-photos-of-2017/#

Kench, P.S., Ford, M.R. and Owen, S.D. (2018) Patterns of island change and persistence offer alternate adaptation pathways for atoll nations. *Nature Communications,* 9, 605.

Kennedy, F. (2022) Bloodied beaches, copper flowers: A choreopolitical analysis of Extinction Rebellion's Red Rebel Brigade. In M. Zebracki and Z.Z. McNeill (Eds), *Politics as public art: The aesthetics of political organizing and social movements,* Routledge, 82–95.

Kilgo, D.K. and Mourão, R.R. (2021) Protest coverage matters: How media framing and visual communication affects support for Black civil rights protests. *Mass Communication and Society,* 24 (4), 576–596.

Kratzer, R.M. and Kratzer, B. (2003) How newspapers decided to run disturbing 9/11 photos. *Newspaper Research Journal* 24(1), 34.

Krause, A. and Bucy, E.P. (2018) Interpreting images of fracking: How visual frames and standing attitudes shape perceptions of environmental risk and economic benefit. *Environmental Communication,* 12(3), 322–343.

Kress, G. and van Leeuwen, T. (2002) Colour as a semiotic mode: Notes for a grammar of colour. *Visual Communication,* 1(3), 343–368.

Kress, G. and van Leeuwen, T.V. (2020) Reading images: The grammar of visual design (3rd ed.). Routledge. https://doi.org/10.4324/9781003099857

Krulwich, R. (2014, 1 March). Polar bear flip-flop: People hated, then loved these photos. What changed? https://www.npr.org/sections/krulwich/2014/03/01/283993033/polar-bear-flip-flop-people-hated-then-loved-these-photos-what-changed.[accessed 3 July 2025]

Lange, D. (1936) Migrant mother: Digital files from original negatives. U.S. Farm Security Administration/Office of War Information. Prints & Photographs Division.

Lang, P.J., Bradley, M.M. and Cuthbert, B.N. (1990) Emotion, attention, and the startle reflex. *Psychological Review*, 97, 377–395.

Law, J. (2002) Objects and spaces. *Theory, Culture & Society*, 19(5–6), 91–105.

Leaver, T., Highfield, T. and Abidin, C. (2020) *Instagram: Visual social media cultures*. Digital Media and Society Series, Polity.

Lehman, B., Thompson, J., Davis, S. and Carlson, J.M. (2019) Affective images of climate change. *Frontiers in Psychology*, 10, 960.

Leiserowitz, A. (2006) Climate change risk perception and policy preferences: The role of affect, imagery, and values. *Climatic Change*, 77(1), 45–72.

Leiserowitz, A.A., Maibach, E.W., Roser-Renouf, C., Smith, N. and Dawson, E. (2013) Climategate, public opinion, and the loss of trust. *American Behavioral Scientist*, 57(6), 818–837.

León, B. and Erviti, M.C. (2013) Science in pictures: Visual representation of climate change in Spain's television news. *Public Understanding of Science*, 24, 183–199. https://doi.org/10.1177/0963662513500196

León, B., Negredo, S. and Erviti, M.C. (2022) Social engagement with climate change: Principles for effective visual representation on social media. *Climate Policy*, 22(8), 976–992.

Lester, L. (2010) *Media and environment: Conflict, politics and the news*. Polity Press.

Lester, L. and Cottle, S. (2009). Visualising climate change: Television news and ecological citizenship. *International Journal of Communication*, 3, 920–936.

Leviston, Z., Price, J. and Bishop, B. (2014) Imagining climate change: The role of implicit associations and affective psychological distancing in climate change responses. *European Journal of Social Psychology*, 44, 441–454.

Lewis, S.C. and Karoly, D.J. (2013) Anthropogenic contributions to Australia's record summer temperatures of 2013. *Geophysiscal Research Letters*, 40, 3705–3709.

Liao, S. and Gendler, T. (2019) Imagination. In E.N. Zalta (Ed), *The Stanford encyclopedia of philosophy*. https://plato.stanford.edu/archives/win2019/entries/imagination [accessed 17 May 2024].

Libarkin J.C., Thomas S.R. and Ruetenik G. (2013) *Visual salience in climate change imagery is in the eye of the beholder*. Eye-tracking Mini-Conference, Lansing, MI, 22–24 April.

Library of Congress (2014) Cinematic treasures named to National Film Registry. 17 December 2014. https://www.loc.gov/item/prn-14-210/new-films-added-to-national-registry/2014-12-17/ [accessed 8 December 2023].

Linder, S.H. (2006) Cashing-in on risk claims: On the for-profit inversion of signifiers for 'global warming'. *Social Semiotics*, 16(1), 103–132. https://doi.org/10.1080/10350330500487927

Lopes, L.S. and Azevedo, J. (2023) The images of climate change over the last 20 years: What has changed in the Portuguese press? *Journalism and Media,* 4, 743–759.

Lorenz, S., Dessai, S., Forster, P.M. and Paavola, J. (2015) Tailoring the visual communication of climate projections for local adaptation practitioners in Germany and the UK. *Philosophical Transactions of the Royal Society A,* 373, 20140457.

Lough, K. and McIntyre, K. (2023) A systematic review of constructive and solutions journalism research. *Journalism*, 24(5), 1069–1088.

Loverock, B. and Hart, M.M. (2018) What a scientist looks like: Portraying gender in the scientific media. *FACETS,* 3, 754–763.

LSE (2022) What is the role of nuclear in the energy mix and in reducing greenhouse gas emissions? Energy and Climate change Explainers. https://www.lse.ac.uk/granthaminstitute/explainers/role-nuclear-power-energy-mix-reducing-greenhouse-gas-emissions/ [accessed 20 October 2023].

Lu, Y. and Peng, Y. (2024) The mobilizing power of visual media across stages of social-mediated protests. *Political Communication*, 41(4), 531–558. doi: 10.1080/10584609.2024.2317951.

Lück, J., Wozniak, A. and Wessler, H. (2016) Networks of coproduction: How journalists and environmental NGOs create common interpretations of the UN climate change conferences. *The International Journal of Press/Politics*, 21(1), 25–47.

Mahony, M. (2015) Climate change and the geographies of objectivity: The case of the IPCC's burning embers diagram. *Transactions of the Institute of British Geographers*, 40: 153–167.

Mahony, M. and Hulme, M. (2012) The colour of risk: an exploration of the IPCC's 'burning embers' diagram. *Spontaneous Generations,* 6, 75–89.

Manzo, K. (2008) Imaging humanitarianism: NGO identity and the iconography of childhood. *Antipode,* 40, 632–657.

Marcinkowski, F. (2014) Mediatisation of politics: Reflections on the state of the concept. *Javnost–The Public*, 21(2), 5–22.

Masters, R.D., Sullivan, D.G., Lanzetta, J.T., McHugo, G.J. and Englis, B.G. (1986) The facial displays of leaders: Toward an ethology of human politics. *Journal of Social and Biological Structures*, 9, 319–343.

Maurice Smith, D. (2022) A year ago, the first of two catastrophic floods hit the Northern Rivers region of NSW, Australia where I live … Instagram post, 28 February 2023. https://www.instagram.com/p/CpL5r_4BYFi/?img_index=1 [accessed 26 July 2023].

McDonald, R.I., Hui, Y.C. and Newell, B.R. (2015) Personal experience and the 'psychological distance' of climate change: An integrative review. *Journal of Environmental Psychology,* 44, 109–118.

McGarry, A., Jenzen, O., Eslen-Ziya, H., Erhart, I. and Korkut, U. (2019) Beyond the iconic protest images: The performance of 'everyday life' on social media during Gezi Park. *Social Movement Studies,* 18(3), 284–304.

McGivern, A. (2022) [Twitter], 17 July 2022. https://x.com/aidanweather/status/1548603406541111297?ref_src=twsrc%5Etfw%7Ctwcamp%5Etweetembed%7Ctwterm%5E1548603406541111297%7Ctwgr%5E%7Ctwcon%5Es1_&ref_url=https%3A%2F%2Fvivo.tools.bbc.co.uk%2F%2Fstream%2Fcom%2FBE51970A68444FD6B6C424F3E7C70110 [accessed 11 December 2023].

McLachlan, C. (2009) 'You don't do a chemistry experiment in your best china': Symbolic interpretations of place and technology in a wave energy case. *Energy Policy,* 37 (12), 5342–5350.

McLoughlin, N. (2020) *Communicating adaptation: Using psychological insights to facilitate adaptive responses to climate change impacts.* PhD thesis. Psychology, University of Bath.

McMahon, R., Stauffacher, M. and Knutti, R. (2016) The scientific veneer of IPCC visuals. *Climatic Change,* 138, 369–381.

McQuire, S. (1997) *Visions of modernity: Representation, memory, time and space in the age of the camera.* Sage.

Mead, M. and Metraux, R. (1957) Image of the scientist among high-school students. *Science,* 126(3270), 384–390.

Messaris, P. and Abraham, L. (2001) The role of images in framing news stories. In S.D. Reese, O.H.J. Gandy and A.E. Grant (Eds), *Framing public life.* Taylor & Francis, 215–226.

Met Office (2023) No need to see red over Met Office colour scale. *Met Office official blog,* 27 July https://blog.metoffice.gov.uk/2023/07/27/no-need-to-see-red-over-met-office-colour-scale/ [accessed 24 November 2023].

Met Office (2023) UK regional climates. https://www.metoffice.gov.uk/research/climate/maps-and-data/regional-climates/index [accessed 23 June 2023].

Met Office (2023) What is a heatwave? https://www.metoffice.gov.uk/weather/learn-about/weather/types-of-weather/temperature/heatwave [accessed 23 June 2023].

Metag, J., Metag, J., Schäfer, M.S., Füchslin, T., Barsuhn, T. and Kleinen-von Königslöw, K. (2016) Perceptions of climate change imagery: Evoked salience and self-efficacy in Germany, Switzerland and Austria. *Science Communication,* 38, 197–227.

MetsUnite (2023) Meteorologists united on climate change. https://twitter.com/MetsUnite [accessed 24 November 2023].

Mezzofiore, G. (2019) The UK just went one week without coal power for the first time in 137 years. *CNN*, 9 May 2019. https://edition.cnn.com/2019/05/09/health/uk-coal-electricity-renewables-health-scli-intl/index.html [accessed 17 November 2023].

Mildenberger, M., Sahn, A., Miljanich, C., Hummel, M.A., Lubell, M. and Marlon, J.R. (2024) Unintended consequences of using maps to communicate sea-level rise. *Nature Sustainability*. https://doi.org/10.1038/s41893-024-01380-0

Milman, O., Watts, J. and Phillips, T. (2017) Worried world urges Trump not to pull out of Paris climate agreement. *The Guardian*, 7 May 2017. https://www.theguardian.com/environment/2017/may/07/trump-climate-change-officials-worried [accessed 23 June 2023].

Mirzoeff, N. (2011) *The right to look: A counterhistory of visuality*. Duke University Press.

Mittermeier, C.G. (2018, August issue). Starving-polar-bear photographer recalls what went wrong. https://www.nationalgeographic.com/magazine/2018/08/explore-through-the-lens-starving-polar-bear-photo/ [accessed 3 July 2025].

Molder, A.L., Lakind, A., Clemmons, Z.E. and Chen, K. (2021) Framing the global youth climate movement: A qualitative content analysis of Greta Thunberg's moral, hopeful, and motivational framing on Instagram. *The International Journal of Press/Politics*, 27(3), 668–695.

Mookherjee, N. (2018) Memory. In R. Bleiker (Ed), *Visual global politics*. Routledge.

Moreau, M. and Mendick, H. (2012) Discourses of women scientists in online media: Towards new gender regimes? *International Journal of Gender, Science and Technology*, 4(1), 4–23.

Morelli, A., Johansen, T.G., Pidcock, R., Harold, J., Pirani, A., Gomis, M., Lorenzoni, I., Haughey, E. and Coventry, K. (2021) Co-designing engaging and accessible data visualisations: a case study of the IPCC reports. *Climatic Change*, 168, 26.

Moreshead, A. and Salter, A. (2023) Knitting the in_visible: Data-driven craftivism as feminist resistance, *Journal of Gender Studies*, 32(8), 875–886.

Moser, S.C. (2016) Reflections on climate change communication research and practice in the second decade of the 21st century: What more is there to say?. *WIREs Climate Change*, 7, 345–369.

NASA (2024) Climate legacies: Interactive tool (forthcoming at the time of writing). https://svs.gsfc.nasa.gov/webapps/climate-legacies/ [accessed 3 May 2024].

National Museum of Australia (2023) DEFINING MOMENTS: Heatwaves. 6 September 2023. https://www.nma.gov.au/defining-moments/resources/heatwaves [accessed 26 January 2024].

REFERENCES

Nature (2010) Special collections: Climategate. https://www.nature.com/collections/synrzkgmlf [accessed 30 June 2025].

ND-GAIN (2024) ND-GAIN Country Index: Notre Dame Global Adaptation Initiative. https://gain.nd.edu/our-work/country-index/ [accessed 10 May 2023].

Nellemann, C., Verma, R. and Hislop, L. (Eds) (2011) *Women at the frontline of climate change: Gender risks and hopes. A Rapid Response Assessment.* United Nations Environment Programme, GRID-Arendal.

Nerlich, B. and Jaspal, R. (2014) Images of extreme weather: Symbolising human responses to climate change. *Science as Culture,* 23, 253–276.

Netflix (2023) In honor of show your stripes day: A sustainability 'must-watch' list, 21 June. https://about.netflix.com/en/news/show-your-Stripes-day [accessed 24 November 2023].

Newman, N., Fletcher, R., Eddy, K., Robertson, C.T. and Nielsen, R.K. (2023) Digital news report 2023. *Reuters Institute for the Study of Journalism.*

Nieubuurt, J.T. (2021) Internet memes: Leaflet propaganda of the digital age. *Frontiers in Communication,* 5, 547065.

Nisbet, M.C. and Huge, M. (2006) Attention cycles and frames in the plant biotechnology debate: Managing power and participation through the press/policy connection. *Harvard International Journal of Press/Politics,* 11(2), 3–40. https://journals.sagepub.com/doi/10.1177/1081180X06286701

Nisbet, M.C. and Newman, T.P. (2015) Framing, the media, and environmental communication. In H. Anders and R. Cox (Eds), *The Routledge handbook of environment and communication.* Routledge, 325–338.

Nocke, T., Sterzel, T., Böttinger, M. and Wrobel, M. (2008) Visualization of climate and climate change data: an overview. In *Digital Earth Summit on Geoinformatics 2008: Tools for Global Change Research,* 226–232. https://doi.org/ISBN 978-3-87907-486-0

Notz, D. and SIMIP Community (2020). Arctic sea ice in CMIP6. *Geophysical Research Letters,* 47, e2019GL086749. https://doi.org/10.1029/2019GL086749

O'Neill, B.C. (2023) Envisioning a future with climate change. *Nature Climate Change,* 13, 874–876.

O'Neill, S. (2013) Image matters: Climate change imagery in US, UK and Australian newspapers. *Geoforum,* 49, 10–19.

O'Neill, S. (2019) Guest post: How heatwave images in the media can better represent climate risks. *Carbon Brief,* 29 August 2019. https://www.carbonbrief.org/guest-post-how-heatwave-images-in-the-media-can-better-represent-climate-risks/ [accessed 23 June 2023].

O'Neill, S. (2020) More than meets the eye: A longitudinal analysis of climate change imagery in the print media. *Climatic Change,* 163, 9–26, doi.org/10.1007/s10584-019-02504-8.

O'Neill, S. (2022) Defining a visual metonym: A hauntological study of polar bear imagery in climate communication. *Transactions of the Institute of British Geographers,* 47(4), 1104–1119. doi:10.1111/tran.12543.

O'Neill, S. and Boykoff, M. (2010) *Climate denier, skeptic, or contrarian?* Proceedings of the National Academy of Sciences of the United States of America 107, 39.

O'Neill, S. and Hulme, M. (2009) An iconic approach for representing climate change. *Global Environmental Change,* 19, 402–410.

O'Neill, S. and Nicholson-Cole, S. (2009) Fear won't do it: Promoting positive engagement with climate change through imagery and icons. *Science Communication,* 30, 355–379.

O'Neill, S., Boykoff, M., Day, S. and Niemeyer, S. (2013) On the use of imagery for climate change engagement. *Global Environmental Change,* 23, 413–421.

O'Neill, S., Osborn, T.J., Hulme, M., Lorenzoni, I. and Watkinson, A.R. (2008) Using expert knowledge to assess uncertainties in future polar bear populations under climate change. *Journal of Applied Ecology,* 45(6), 1649–1659

O'Neill, S., Williams, H.T.P, Kurz, T., Wiersma, B. and Boykoff, M. (2015) Dominant frames in legacy and social media coverage of the IPCC Fifth Assessment Report. *Nature Climate Change,* 5, 380–385.

O'Neill, S., Hayes, S., Strauss, N., Doutreix, M., Steentjes, K., Ettinger, J., Westwood, N. and Painter, J. (2023) Visual portrayals of fun in the sun misrepresent heatwave risks in European newspapers. *The Geographical Journal,* 189, 90–103.

Omidi, M. (2009) Maldives sends climate SOS with undersea cabinet. *Reuters,* 19 October. https://www.reuters.com/article/us-maldives-environment-idUSTRE59G0P120091017/ [accessed 3 January 2024].

ONS (2022) Employment by occupation. Office for National Statistics. https://www.ethnicity-facts-figures.service.gov.uk/work-pay-and-benefits/employment/employment-by-occupation/latest [accessed 23 June 2023].

Open Planet (2024) Open Planet: World class footage of our changing planet (website). https://openplanet.org/ [accessed 30 June 2025].

Palmer, R., Toff, B. and Nielsen, R.K. (2020) 'The media covers up a lot of things': Watchdog ideals meet folk theories of journalism. *Journalism Studies*, 21(14), 1973–1989.

Paik, S., Bonna, S., Novozhilova, E., Gao, G., Kim, J., Wijaya, D. and Betke, M. (2023) The affective nature of AI-generated news images: Impact on visual journalism. In S. Canavan and D. Jiang (Eds), 2023 11th International Conference on Affective Computing and Intelligent Interaction, ACII 2023 IEEE, Institute of Electrical and Electronics Engineers. https://doi.org/10.1109/ACII59096.2023.10388166

Panofsky, E. (1970) *Meaning in the visual arts*. Penguin.

Parry, K. (2010) A visual framing analysis of British press photography during the 2006 Israel- Lebanon conflict. *Media, War & Conflict*, 3, 67–68.

Parry, K. (2011) Images of liberation? Visual framing, humanitarianism and British press photography during the 2003 Iraq invasion. *Media, Culture & Society*, 33(8), 1185–1201.

Pastoureau, M. (2016) *Red: The history of a colour*. Princeton University Press.

Patel, K. (2023) This visual shows how climate change will affect generations. *21 March, Washington Post*. https://www.washingtonpost.com/climate-environment/2023/03/21/climate-ipcc-report-temperatures-graphic/ [accessed 24 November 2023].

Pearce, W. and De Gaetano, C. (2021) Google Images, climate change, and the disappearance of humans. *Diseña*, 19(3), ISSN 0718–8447.

Pearce, W. and Nerlich, N. (2018). 'An inconvenient truth': A social representation of scientific expertise. In N. Nerlich et al (Eds), *Science and the Politics of Openness*. Manchester University Press, 212–229.

Pearce, W., Özkula, S.M., Greene, A.K., Teeling, L., Bansard, J.S., Omena, J.J. and Rabello, E.T. (2018) Visual cross-platform analysis: Digital methods to research social media images. *Information, Communication and Society*, 23(2), 161–180.

Pearson, A. (2023) First it was Covid – now we're being scared into submission over the weather. *The Telegraph*, 18 July. https://www.telegraph.co.uk/columnists/2023/07/18/heatwave-europe-scared-by-weather-covid-nudge-unit/ [accessed 24 November 2023].

Perlmutter, D.D. (2006). Hypericons: Famous news images in the internet-digital-satellite age. In P. Messaris and L. Humphreys (Eds), *Media: Transformations in human communication*. Peter Lang, pp 51–64.

Petras (2024) *Electric feels: Artistic responses to research exploring human dimensions of digital energy transformation*. https://petras-iot.org/update/electric-feels/ [accessed 26 January 2024].

Petre, C. (2015) The traffic factories: Metrics at Chartbeat, Gawker Media, and The New York Times. *Tow Center for Digital Journalism*. https://www.cjr.org/tow_center_reports/the_traffic_factories_metrics_at_chartbeat_gawker_media_and_the_new_york_times.php [accessed 30 June 2025].

Picfair (2023) Enjoy the rain. Photograph by Muhammad Amdad Hossain, Bangladesh. https://www.picfair.com/pics/09816538-enjoy-the-rain [accessed 26 July 2023].

Plumb, G. (2023) *Show your Stripes 2023, 20 June, Royal Meteorological Society*. https://www.rmets.org/metmatters/show-your-Stripes-2023 [accessed 24 November 2023].

Popp, R.K. and Mendelson, A.L. (2010) 'X'-ing out enemies: time magazine, visual discourse, and the war in Iraq. *Journalism*, 11(2), 203–221.

Preece, R. (2012) How wind farm developers 'use camera tricks to make turbines look smaller than they really are'. *Daily Mail*, 22 August 2012.

Price, V. and Tewksbury, D. (1997) News values and public opinion: A theoretical account of media priming and framing. In G.A. Barett and F.J. Boster (Eds), *Progress in communication sciences: Advances in persuasion*, Vol 13. Ablex, 173–212.

Querubín, N.S. and Niederer, S. (2022) Climate futures: Machine learning from cli-fi. *Convergence*, 0(0).

Reading University (2022) #ShowYourStripes 2022: Join global climate change movement. 20 June. https://www.reading.ac.uk/news/2022/University-News/Countdown-to-Show-Your-Stripes-Day-2022 [accessed 24 November 2023].

Rebanks, J. (2020) *English pastoral: An inheritance*. Penguin Random House.

Rebich-Hespanha, S. (2011). Thematic and affective content in textual and visual communication about climate change: Historical overview of mass media sources and empirical investigation of emotional responses. Unpublished PhD thesis. University of California. http://search.proquest.com/docview/908613624 [accessed 20 October 2023].

Rebich-Hespanha, S. and Rice, R.E. (2016). Dominant visual frames in climate change news stories: Implications for formative evaluation in climate change campaigns. *International Journal of Communication*, 10, 4830–4862.

Rebich-Hespanha, S., Rice, R., Montello, D.R., Retzloff, S., Tien, S. and Hespanha, J.P. (2015) Image themes and frames in US print news stories about climate change. *Environmental Communication*, 9, 491–519.

Remillard, C. (2011) Picturing environmental risk: The Canadian oil sands and the National Geographic. *International Communication Gazette*, 73, 127–143.

Reque, J., Tantillo, S.H., Babb, J., McIntosh, M. and Denham, B. (2001). *Introduction to Journalism*. McDougall Littell.

Reuters (2020) UK greenhouse gas emissions fell 3.6 per cent in 2019. India Times 26 March 2020. https://energy.economictimes.indiatimes.com/news/coal/uk-greenhouse-gas-emissions-fell-3-6-per-cent-in-2019/74828144?redirect=1 [accessed 17 November 2023].

Reuters Institute (2024) *AI and the future of news, Reuters Institute for the Study of Journalism, University of Oxford*, https://reutersinstitute.politics.ox.ac.uk/ai-journalism-future-news [accessed 17 April 2024].

Richardson, M. (2023) *#BiodiversityStripes*, University of Derby. https://biodiversityStripes.info/global/ [accessed 24 November 2023].

Ribbans, E. (2020) A photograph that is right for this website can be wrong for social media. *The Guardian*, 16 February. https://www.theguardian.com/commentisfree/2020/feb/16/images-death-distress-photograph-publish-social-media [accessed 30 June 2025].

Roe, S. (2023) The 'Climate Generations' figure from @IPCC_CH SYR is one of the most compelling dataviz examples I have come across. 25 May. https://twitter.com/stephanieroe/status/1661723306544750593 [accessed 24 November 2023].

Romanello, M. et al (2022) The 2022 report of the Lancet Countdown on health and climate change: health at the of fossil fuels. *Lancet*, 400, 1619–1654.

Rose, G. (2016) Rethinking the geographies of cultural 'objects' through digital technologies: Interface, network and friction. *Progress in Human Geography*, 40(3), 334–351.

Rose, G. (2023) *Visual Methodologies: An introduction to researching with visual materials*, 5th ed. Sage.

Ross, A.S. and Rivers, D.J. (2017) Digital cultures of political participation: Internet memes and the discursive delegitimization of the 2016 U.S. Presidential candidates. *Discourse, Context and Media*, 16, 1–11.

Ross, A.S. and Rivers, D.J. (2019) Internet memes, media frames, and the conflicting logics of climate change discourse, *Environmental Communication*, 13(7), 975–994.

Royal Society (2023) Five graphs that changed the world – with Adam Rutherford, *The Royal Society*. 24 March. https://www.youtube.com/watch?v=8wtmCXgNqG4 [accessed 24 November 2023].

Russell, S. (2021) Lecture 2: The future role of AI in warfare. BBC Reith Lectures 2021 – Living with Artificial Intelligence. *BBC Radio* 4, Manchester. https://www.bbc.co.uk/programmes/articles/1N0w5NcK27Tt041LPVLZ51k/reith-lectures-2021-living-with-artificial-intelligence [accessed 10 January 2024].

Sakellari, M. (2021) Climate change and migration in the UK news media: How the story is told. *International Communication Gazette*, 83(1), 63–80.

Scarles, C. (2004) Mediating landscapes: The processes and practices of image construction in tourist brochures of Scotland. *Tourist Studies*, 4(1), 43–67. https://doi.org/10.1177/1468797604053078

Schäfer, M.S., Ivanova, A. and Schmidt, A. (2014) What drives media attention for climate change? Explaining issue attention in Australian, German and Indian print media from 1996 to 2010. *International Communication Gazette*, 76, 152–176.

Schäfer, M. and O'Neill, S. (2017) *Frame analysis in climate change communication. The Oxford Encyclopaedia of climate change communication.* Oxford Research Encyclopaedia (Climate Science), Oxford University Press.

Scheufele, D.A. and Nisbet, M. (2007) Framing. *In* L.L. Kaid, and C. Holtz-Bacha (Eds), *Encyclopaedia of Political Communication*. Sage.

Schneider, B. (2016) Burning worlds of cartography: A critical approach to climate cosmograms of the Anthropocene. *Geo*, 3(2), e00027.

Schneider, B. and Nocke, T. (Eds) (2014) *Image politics of climate change: Visualizations, imaginations, documentations*. Transcript Verlag.

Schneider, B. and Nocke, T. (2018) The feeling of red and blue: A constructive critique of color mapping in visual climate change communication. In B. Schneider and T. Nocke (Eds), *Image politics of climate change*. Transcript Verlag.

Schneider, J. (2021) Getty photographer shares a look back at 2021 in photos. *PetaPixel*, 17 December 2021. https://petapixel.com/2021/12/17/getty-photographer-shares-a-look-back-at-2021-in-photos/

Schneider, S. (2001) What is 'dangerous' climate change? *Nature*, 411, 17–19.

Schwab, K. (2019) Crafting takes a dark turn in the age of climate crisis. 1 November 2019. https://www.fastcompany.com/90290800/crafting-take-a-dark-turn-in-the-age-of-climate-crisis [accessed 24 November 2023].

Science & Technology Committee (2010) Government response to the House of Commons Science and Technology Committee 8th Report of Session 2009–10: The disclosure of climate data from the Climatic Research Unit at the University of East Anglia. *UK Government*. https://assets.publishing.service.gov.uk/government/uploads/system/uploads/attachment_data/file/228975/7934.pdf [accessed 26 July 2023].

Seelig, M. (2005) A case for the visual elite. *Visual Communication Quarterly*, 12(3–4), 164–181.

Shah, P. and Hoeffner, J. (2002) Review of graph comprehension research: Implications for instruction. *Educational Psychology Review*, 14(1), 47–69.

Shanahan, M. and Bahia, K. (2023) State of mobile internet connectivity report 2023. October 2023, Global System for Mobile Communications Association (GSMA). https://www.gsma.com/r/somic/ [accessed 27 May 2024].

Sheldon, J. (2017) The globally warm scarf. 8 February. https://sheldonfiberdesigns.net/the-globally-warm-scarf/ [accessed 24 November 2023].

Shenton, J.E. (2020) Divided we tweet: The social media poetics of public online shaming. *Cultural Dynamics*, 32(3), 170–195.

Shields, F. (2019) Why we're rethinking the images we use for our climate journalism. *The Guardian*, 18 October 2019. https://www.theguardian.com/environment/2019/oct/18/guardian-climate-pledge-2019-images-pictures-guidelines [accessed 23 June 2023].

Shifman, L. (2013) Memes in a digital world: Reconciling with a conceptual troublemaker. *Journal of Computer-Mediated Communication*, 18(3), 362–377.

Shutterstock (2024) Shutterstock website https://www.shutterstock.com/ [accessed 27 May 2024].

Silva, M. (2023) False claims that heatwave is bogus spread online. *BBC News*, 28 July. https://www.bbc.co.uk/news/science-environment-66314338 [accessed 24 November 2023].

Simonsen, A.H. (2022) Blowing in the wind: Norwegian wind power photographs in transition. *Journalism Practice*, 16(2–3), 298–316.

SJN (2023) *Solutions Journalism Network Mission.* https://www.solutionsjournalism.org/about [accessed 17 November 2023].

Slezak, M. (2017) Fiji told it must spend billions to adapt to climate change. *The Guardian*, 9 November 2017. https://www.theguardian.com/environment/2017/nov/10/fiji-told-it-must-spend-billions-to-adapt-to-climate-change#:~:text=To%20prepare%20for%20the%20rising,nation's%20vulnerability%20to%20climate%20change%2C [accessed 23 June 2023].

Smith, J.B., Schneider, S.H., Oppenheimer, M., Yohe, G.W., Hare, W., Mastrandrea, M.D., Patwardhan, A., Burton, I., Corfee-Morlot, J., Magadza, C.H., Füssel, H.M., Pittock, A.B., Rahman, A., Suarez, A. and van Ypersele, J.P. (2009) Assessing dangerous climate change through an update of the Intergovernmental Panel on Climate Change (IPCC) 'reasons for concern'. *Proceedings of the National Academy of Sciences USA*, 106(11), 4133–4137.

Smith, N. and Joffe, H. (2009) Climate change in the British press: The role of the visual. *Journal of Risk Research,* 12, 647–663.

Solutions Journalism Network (2024) Responses to problems are newsworthy. https://www.solutionsjournalism.org/ [accessed 18 April 2024].

Sontag, S. (1977) *On photography*. Penguin.

Sparrow, J. (2019) Eco-fascists and the ugly fight for 'our way of life' as the environment disintegrates. *The Guardian*, 29 November 2019. https://www.theguardian.com/environment/2019/nov/30/eco-fascists-and-the-ugly-fight-for-our-way-of-life-as-the-environment-disintegrates [accessed 23 June 2023].

Spiegel, S.J. (2020) Visual storytelling and socioenvironmental change: Images, photographic encounters, and knowledge construction in resource frontiers. *Annals of the American Association of Geographers,* 110(1), 120–144.

Stanford, C. (2023) The 15-minute city: Where urban planning meets conspiracy theories. *New York Times*, 1 March 2023, https://www.nytimes.com/2023/03/01/world/europe/15-minute-city-conspiracy.html [accessed 20 October 2023].

Strauss, D.L. (2020) On images & magic: Towards an iconopolitics of belief. In P.L. Wilson et al (Eds), *The critique of the image is the defense of the imagination*, Autonomedia.

Strauss, D.L. (2020) *Photography and belief*, David Zwirner Books.

Strauss, N., Painter, J., Ettinger, J., Doutreix, M.N., Wonneberger, A. and Walton, P. (2021) Reporting on the 2019 European heatwaves and climate change: Journalists' attitudes, motivations and role perceptions. *Journalism Practice*, 16, 462–485. https://doi.org/10.1080/17512786.2021.1969988

Strengers, Y. and Kennedy, J. (2020) *The smart wife: Why Siri, Alexa, and Other smart home devices need a feminist reboot*. MIT Press.

Takach, G. (2013) Selling nature in a resource-based economy: Romantic/extractive gazes and Alberta's bituminous sands. *Environmental communication*, 7(2), 211–230.

Takacs, B. and Goulden, M.C. (2019) Accuracy of wind farm visualisations: The effect of focal length on perceived accuracy. *Environmental Impact Assessment Review*, 76, 1–9.

Takahashi, B., Metag, J., Thaker, J. and Evans Comfort, S. (Eds) (2021) *The handbook of international trends in environmental communication*. Routledge.

TalkTV (2023) Julia blasts the 'hysterical' reporting of the European heatwave suggesting 'humanity is going to end', 18 July. https://twitter.com/TalkTV/status/1681200908082769928?s=20 [accessed 24 November 2023].

Tamble, M. (2019) 7 tips for using visual content marketing, 20 February 2019. https://www.socialmediatoday.com/news/7-tips-for-using-visual-content-marketing/548660/#:~:text=The%20click%2Dthrough%2Drate%20(,than%20content%20without%20relevant%20images [accessed 12 April 2024].

Tamman, M. (2021) The hot list, meet Julie Arblaster. 26 April. https://www.reuters.com/investigates/special-report/climate-change-scientists-arblaster/ [accessed 24 November 2023].

Taylor, A. (2017) 2017 seen through the lens of Mario Tama. *The Atlantic*, 27 December 2017. https://www.theatlantic.com/photo/2017/12/2017-seen-through-the-lens-of-mario-tama/549266/ [accessed 23 June 2023].

TCAP (2023) Tuvalu coastal adaptation project Facebook page. https://www.facebook.com/TuvaluCoastalAdaptationProject [accessed 23 June 2023].

Text2Data (2023) *Sentiment analysis* https://text2data.com/Demo [accessed 17 Nov 2023].

The Economist (2019) The climate issue. 9 September. https://www.economist.com/leaders/2019/09/19/the-climate-issue [accessed 24 November 2023].

Their, K. and Lin, T. (2022) How solutions journalism shapes support for collective climate change adaptation. *Environmental Communication*, 16(8), 1027–1045.

Thomas, R.J. and Thomson, T.J. (2023) What does a journalist look like? Visualizing journalistic roles through AI. *Digital Journalism*. https://doi.org/10.1080/21670811.2023.2229883

Thompson, C.E. (2019) 'We are here': New climate design shows us our future in red-hot Stripes. Grist, 21 March 2019. https://grist.org/climate-energy/we-are-here-new-climate-design-shows-us-our-future-in-red-hot-Stripes/ [accessed 26 January 2024].

Thomson, T. (2024) Picturing a quality local news visual: Perspectives from non-specialist journalists. *Journalism*, 0(0). https://doi.org/10.1177/14648849241253136

Thomson, T.J. and Greenwood, K. (2017). I 'like' that: Exploring the characteristics that promote social media engagement with news photographs. *Visual Communication Quarterly*, 24(4), 203–218.

REFERENCES

Thomson, T.J., Angus, D., Dootson, P., Hurcombe, E. and Smith, A. (2022) Visual mis/disinformation in journalism and public communications: current verification practices, challenges, and future opportunities. *Journalism Practice*, 16(5), 938–962.

Thunberg, G. (2022) *The climate book*. Allen Lane.

Titifanue, J., Kant, R., Finau, G. and Tarai, J. (2017) Climate change advocacy in the Pacific: The role of information and communication technologies. *Pacific Journalism Review*, 23, 133–149.

Treen, K., Williams, H., O'Neill, S. and Coan, T. (2022) Discussion of climate change on Reddit: Polarised discourse or deliberative debate? *Environmental Communication*, 16(5). https://doi.org/10.1080/17524032.2022.2050776

Tukachinsky, R., Mastro, D. and King, A. (2011). Is a picture worth a thousand words? The effect of race-related visual and verbal exemplars on attitudes and support for social policies. *Mass Communication and Society*, 14(6), 720–742.

UBC (2023) Live: Good Morning Uganda. UBC Television Uganda. https://www.youtube.com/watch?v=kBMRwLRj4Nk [accessed 24 November 2023].

UEA (2021) The story behind BBC drama The Trick: The truth will set you free. https://stories.uea.ac.uk/the-story-behind-the-trick/ [accessed 15 March 2024].

UEA (2023) A climate mural for our times: Global science, local expression. https://stories.uea.ac.uk/a-climate-mural-for-our-times/ [accessed 24 November 2023].

UKHSA/ONS (2022) Heat mortality monitoring report: 2022. https://www.gov.uk/government/publications/heat-mortality-monitoring-reports/heat-mortality-monitoring-report-2022#:~:text=During%20summer%202022%2C%20there%20were,number%20in%20any%20given%20year [accessed 23 June 2023].

UN (1992) United Nations Framework Convention on Climate Change. FCCC/INFORMAL/84 GE.05-62220 (E) 200705.

UNFCCC (2023) The Paris Agreement: What is the Paris Agreement? United Nations Framework Convention on Climate Change. https://unfccc.int/process-and-meetings/the-paris-agreement#:~:text=It%20entered%20into%20force%20on,above%20pre%2Dindustrial%20levels.%E2%80%9D [accessed 24 November 2023].

UNGeneva (2023) Over 2 M deaths and $4.3 trillion in economic losses; that's the impact of a half-century of extreme weather events. *Tweet*, 23 May 2023. https://twitter.com/UNGeneva/status/1660986686007029763 [accessed 8 March 2024].

UNICEF (2024) https://www.unicef.org/goodwill-ambassadors/vanessa-nakate#:~:text=Vanessa%20began%20advocating%20for%20climate,%E2%80%9Cstriking%E2%80%9D%20for%20the%20climate [accessed 30 June 2025].

United States Department of State (2021) *The long-term strategy of the United States: pathways to net-zero greenhouse gas emissions by 2050.* United States Department of State and the United States Executive Office of the President, Washington DC. November 2021. https://www.whitehouse.gov/wp-content/uploads/2021/10/US-Long-Term-Strategy.pdf [accessed 20 October 2021].

University of York (2022) *Show Your Stripes at Pride 2022.* Department of Chemistry, University of York https://www.york.ac.uk/chemistry/news/deptnews/2022/show-your-Stripes/ [accessed 24 November 2023].

Urry, J. (1992) The tourist gaze 'revisited'. *American Behavioural Scientist,* 36(2), 172–186.

Usery, A.G. (2022) Solutions journalism: How its evolving definition, practice and perceived impact affects underrepresented communities. *Journalism Practice,* 18(8), 1887–1903.

V&A (2023) Electric feels: Exhibition at the Victoria and Albert Museum, 23–24 September 2023. https://www.vam.ac.uk/event/LQDLP6mWbr/electric-feels [accessed 20 October 2023].

van der Linden, S.L., Leiserowitz, A.A., Feinberg, G.D. and Maibach, E.W. (2015) The scientific consensus on climate change as a gateway belief: experimental evidence. *PLoS ONE,* 10(2), e0118489.

van Leeuwen, T. (2004) Ten reasons why linguists should pay attention to visual communication. In P. LeVine and R. Scollon (Eds), *Discourse and technology: multimodal discourse analysis.* Georgetown University Press, 7–19.

van Zanten, B.T., Verburg, P.H., Koetse, M.J. and van Beukering, P.J.H. (2014) Preferences for European agrarian landscapes: A meta-analysis of case studies. *Landscape and Urban Planning,* 132*, 89–101.*

Verchot, M. and Biswas, S. (2024) Introducing the capturing climate change cohort, 14 June 2024. https://inoldnews.com/introducing-the-capturing-climate-change-team/ [accessed 5 July 2024].

Veneti, A. and Rovisco, M. (2023) *Visual politics in the Global South.* Palgrave Macmillan.

Vespa, M., Schweizer-Ries, P., Hildebrand, J. and Kortsch, T. (2022) Getting emotional or cognitive on social media? Analyzing renewable energy technologies in Instagram posts. *Energy Research & Social Science,* 88, 102631.

Vevea, N.N., Littlefield, R.S., Fudge, J. and Weber, A.J. (2011). Portrayals of dominance: Local newspaper coverage of a natural disaster. *Visual Communication Quarterly,* 18(2), 84–99. https://doi.org/10.1080/15551393.2011.574064

Victor, D. and Kennel, C. (2014) Ditch the 2°C warming goal. *Nature*, 514, 30–31.

Vlasceanu, M. et al (2024) Addressing climate change with behavioral science: A global intervention tournament in 63 countries. *Science Advances*, 10(6). https://ora.ox.ac.uk/objects/uuid:b80cb5ea-5546-4f6c-902f-83be24731c76

Weaver, I., Westwood, N., Coan, T., O'Neill, S. and Williams, H.T.P. (2022) Sponsored messaging about climate change on Facebook: Actors, content, frames. *arXiv pre-print*, 25 November 2022. https://doi.org/10.48550/arXiv.2211.13965

Wessler, H., Wozniak, A., Hofer, L. and Lück, J. (2016) Global multimodal news frames on climate change: A comparison of five democracies around the world. *The International Journal of Press/Politics*, 21(4), 423–445.

Westwood, N. (2024) *Advertising the heroes and villains of climate change: Narratives in sponsored posts on Facebook and Instagram*. PhD thesis, University of Exeter.

White, D. (2022) Keeping the peace: The visual in the 'struggle' of non-violent activism in a global existential crisis. In S. Hartle and D. White (Eds), *Visual activism in the 21st century: Art, protest and resistance in an uncertain world*. Bloomsbury Visual Arts, Bloomsbury Collections.

Whitmore, G. (2018) Observer picture archive: The standalone photograph, 25 August 2018. https://www.theguardian.com/artanddesign/2018/aug/25/observer-archive-the-standalone-photograph [accessed 11 April 2024].

WHO (2018). Heat and health. *World Health Organisation*. https://www.who.int/news-room/fact-sheets/detail/climatechange-heat-and-health [accessed 21 March 23].

WHO (2024) Health topics: heatwaves. https://www.who.int/health-topics/heatwaves#tab=tab_1 [accessed 10 January 2024].

Wiersma, B. and Devine-Wright, P. (2014) Public engagement with offshore renewable energy: a critical review. *WIREs: Climate Change,* 5, 493–507.

Wikipedia (2023) United States House Select Committee on the climate crisis. https://en.wikipedia.org/wiki/United_States_House_Select_Committee_on_the_Climate_Crisis [accessed 24 November 2023].

WMO (2024) Topics: Climate https://wmo.int/topics/climate [accessed 10 January 2024].

Wolf, J., Adger, W.M. and Lorenzoni, I. (2010) Heat waves and cold spells: An analysis of policy response and perceptions of vulnerable populations in the UK. *Environment and Planning A*, 42, 2721–2734.

World Press Photo (2018) *The state of news photography 2018: Photojournalists' attitudes toward work practices, technology and life in the digital age*. Industry report, edited by A. Hadland and C. Barnett. https://www.worldpressphoto.org/getmedia/4f811d9d-ebc7-4b0b-a417-f119f6c49a15/the_state_of_news_photography_2018.pdf [accessed 26 October 2023].

World Weather Attribution (2024) World Weather Attribution website. https://www.worldweatherattribution.org/about/ [accessed 3 July 2025].

Wozniak, A. (2021) Just 'performance nonsense'? How recipients process news photos of activists' symbolic actions about climate change politics. *Nordic Journal of Media Studies*, 3(1), 61–78.

Wozniak, A., Wessler, H. and Lück, J. (2017) Who prevails in the visual framing contest about the United Nations climate change conferences? *Journalism Studies*, 18(11), 1433–1452.

Yale Program on Climate Change Communication (2023) Global warming's six Americas. https://climatecommunication.yale.edu/about/projects/global-warmings-six-americas/ [accessed 24 November 2023].

Yu, H. and Chen, G. (2021). Their floods and our floods: News values of flood photo galleries of Associated Press and Xinhua News Agency. *Journalism*, 24 (6), 1362–1381. https://doi.org/10.1177/14648849211056785

Zhang, B. and Pinto, J. (2021) Changing the world one meme at a time: The effects of climate change memes on civic engagement intentions. *Environmental Communication*, 15(6), 749–764.

Zhao, X., Jackson, D. and Nguyen, A. (2022) The psychological empowerment potential of solutions journalism: Perspectives from pandemic news users in the UK. *Journalism Studies*, 23(3), 356–373.

Zhong, S., Yang, L., Toloo, S., Wang, Z., Tong, S., Sun, X., Crompton, D., FitzGerald, G. and Huang, C. (2018) The long-term physical and psychological health impacts of flooding: A systematic mapping. *Science of the Total Environment*, 626, 165–194.

Zoch, L.M. and Turk, J.V.S. (1998) Women making news: Gender as a variable in source selection and use. *Journal of Mass Communication Quarterly*, 75(4), 762–775.

Zuckerman, E. (2021) Why study media ecosystems? *Information, Communication & Society*, 24(10), 1495–1513.

Index

References to figures and photographs appear in *italic* type.

15 Minute City 60

A

abstract climate change images 82
Ackerley, Duncan 67, *68*
'action-based storytelling' (De Meyer) 58
adaptation 11–28
  cultural barriers 44
  defining 11
  to extreme heat 15
  to heatwaves 11–12
  small islands 18–19
  urban areas 15
  women 44–45
adaptation imagery 27–28
Adolphsen, M. 88
aerial images 19–21, *20*, 25, 109
'affective imagery' 107–108
affordances of colour 74
African-Americans as helpless 44, 45
Agence France-Press 106
The Age (newspaper) 85, *86*
Aiello, Giorgia 49, 106
Algemeen Dagblad (AD) 15
Ali, Z.S. 44, 45
alternative heatwave imagery 16, *16*
Amdad Hossain, Muhammad 43
AMOS website 75
Anderson, C.W. 112
anthropogenic climate change 48
Apollo 11 'blue marble' images 54
applicability model (Price and Tewksbury) 5
Arblaster, Julie 74–75
Arctic 31
  Canadian 36
article texts 13, 14
artificial intelligence (AI) 2, 9, 121–122
Associated Press 106
audience research 81, 89, 113

Australia
  'angry summer' heatwave 75, 77
  Bureau of Meteorology 74–75, *76*
  climate politics 85, *86*
  'fun in the sun' visual frame 17
  Green Party 85
  purple colour chart 113
  weather maps 74–75, *76*, 116
Auth, Tony (William Auth) 47
Axios 73

B

Barnett, J. 19, 25
BBC 16, 51, 78–79, 113, 116
  Climate Disinformation Reporter 79–80
Ben-Ari, Rafael *21*, 22
Bennet, C., 45
Bennett, W.L. 84
Beradelli, Jeff 70
Berger, John 9
Berlin Zoo 32
'Better Images of AI' (Dihal and Duarte) 2
Biden, Joseph (Joe) 98
Birt, Arlene 73
Bissell, Kimberly 104
Bleiker, Roland 4, 9
Born, D. 30
Bourne, Joe 58–59
Boussalis, Constantine 16
Brendel, Kat 91
Briggs, Laura 1–2
Brown, Bob 85–86, *86*
Buckland, Ella and Myla *40*, 41, 42, 103
Bucy, E.P. 60, 85
Bulivono, Miriam 21–25, *21*, 107, 109
'Burning Embers' diagram (IPCC) 62, 63–66, *64*, 115–116
Bush, George W. 32
Butfield, Colin 119

157

C

Campbell, David 2
Cann, Tristan 49, 52
Caple, Helen 48
Capturing Climate Change project 119
Carbon Brief 12, 56–57
Carr, Esyllt 113
Carter, Jimmy 47
Catto, Jennifer 67, *68*
celebrities 84
Chadwick, A. 101
Chambers, D.W. 1
Chittagong, Bangladesh 43
Clerwall, C. 113
Climate Book (Thunberg) 70
climate change
  defining 6
  image problem 123
  mass protests 35, 93–95
  online platform aesthetics 115
  'sinking islands' visual trope 25
  studies of news stories 4, 16–17
  visual representations 7
  visual tropes 25
'Climate change and the integrity of science' (Gleick) 35
The Climate Collection 52
climate controversies 74, 75
climate-energy case studies 58
'Climategate' 32–33, *34*, 86
Climate Generations figure (IPCC AR6 Synthesis Report) 71, 72–73, 74, 80
climate impacts 29, 38, 44–45
Climate Leadership Corps 32
climate mitigation 51, 60
climate news 23–24, 84–86, 115
Climate Outreach 117, 119
climate protests
  mass protests 35, 93–95
  performance protests 90–93, *90*
climate reporting 106
climate scepticism 32
climate science visuals 72, 62–82
climate solutions journalism 15, 118
Climate Spiral (Hawkins, Fæhn and Fuglestvedt) 62, 66–67, 74, 80, 82
Climate Stripes 62, *70, 71, 72,* 74, 80–81, 82, 112
Climate Stripes blanket (Highwood) 67–68, *68*
climate visual communication 8, 116–122
climate visual discourses 23, *23*, 116–117, 122–124
Climate Visuals programme 117, 119
climate vulnerable countries 17
Climate Warriors campaign imagery 26
Climatic Research Unit (CRU) email controversy 32–33, *34*, 86

Clinton, Hillary 95
CNN 50
Coan, Travis 16
Coastal and Estuarine Research Federation 67
Coca-Cola 30
Coleman, Renita 7
Colose, Chris 74
colours
  affordances 74
  cultural norms 51
  interpretation 80
  role of in images 115–116
  and temperature 76, 77–80
colour wheel of visual imagery 87, *87*
  other uses 50, 54, 87
communications cycle 100
composite images 35–36, *36*, 115
'Condescending Wonka' meme 96–97, *96*
Conference of the Parties (COP) 85, 86, 86–89, *88*, 111
  COP15 (Copenhagen) 33, 86, 90, *90*
  COP26 (Glasgow) 24–25, 87–89, *87, 88,* 103, 106
  COP27 (Sharm-El-Sheikh) 86
content management systems 24
  see also image management systems
'convinced' logic memes 96–98, *96*, 114–115
coral atolls 19–21, *20*, 109
Corner, A. 57
Corriere Della Sera 73
Cosgrove, Denis 30–31, 54
Creative Commons Licenses (CC) 112, 119
'creative' images *see* stock images
cropping photos 94, 103–104
'culture of the click' (Anderson) 112
Cyclone Winston *21*, 22, 109

D

Dahwall, Rubert (RD) 79
Daily Express 84, 108–109
Daily Mail 30, 32
  'Greenwash!' article 52
  Smokestacks imagery 48
The Daily Telegraph see The Telegraph
Dal, John *39*
'Damsel in distress' portrayals 45
Darfur 2
'datafied version of the audience' (Dodd) 113
Davoudi, S. 3
De Gaetano, C. 111
della Dora V. 30–31
De Meyer, Kris 58
depersonalised protesters 93
Devine-Wright, H. 51–52
Devine-Wright, Patrick 56
DiFrancesco, D.A. 14, 115
digital interfaces 100

INDEX

digital technologies 3
Dihal, Kanta 2
diversity of visuals 121
Dixon, Deborah 75
documentaries 23
Dodds, T. 113
Donner, Simon 109
doom and gloom messaging 113
Duan, R., 82
Duarte, Tania 2
Dunaway, F. 47
Dunn, Katherine 111, 121

E

Earth Island Institute 13
The Economist 70
editorial images 22, 106
'Electric Feels' (Michalec and Bourne) 58–59
electricity pylons 46
'Ella Buckland and her daughter Myla in Lismore' photo *40*, 41
'emaciated polar bear' images 36–37, 113
email hacks 32–33
emotional portrayals 44
energy and climate mitigation 46–47
energy debates 60
energy futures 56, 57–58
energy imagery 45–61, *57*, *59*, 104–105
Entman, Robert 4
'eschatological' landscapes 31
Extinction Rebellion (XR) 91–92
extreme heat
  behavioural adaptations 15
  impacts on at-risk people 103, 120
  media visual coverage 12–13
  purple colour controversy 75–77, *76*
  uneven impacts 16, *16*
  see also heatwave visuals
eye-tracking technology 81

F

Facebook 95, 114
  Ad Library 97
facial expressions 85–86, *86*
'Facing the Electorate' (Grabe and Bucy) 85
Fæhn, Taran 66–67
'Family lifestyle travel wind energy' collection (Me 3645 Studio) 57–58
famines 1–2, 45
Farage, Nigel 106
Farbotko, Carol 23, *23*, 25–26
'fear won't do it' 113–114
Feldman, L. 56
Financial Times 73
Fischer, H. 81
flooding images 37–45, 118
  emotional portrayals 44
  Fiji 21–24, *21*, 107, 109
  gendered visual representations 43–44
  'getting on with it' imagery 42–43
  Global North 37–38, 43
  Global South 38, 42–43
  human suffering 38–42, 43
  Lismore flooding 39–41, *40*, 103
  Morpeth floods *39*, 102–103
  newsworthy photographs 42
  racial stereotyping 44
  rescues *39*, 102
  visual tropes 42
Foley, M. 45
Fox, K. 12
framing 4–5
Francisco, Doug 91–92
Fraser, Jim 75
front-of-mind visuals *13*, 107
Frosh, Paul 105–106, 106
Fruean, Brianna 24–25
Fuglestvedt, Jan 66–67
Funafuti, Tuvalu 19–21
'fun in the sun' visual frame
  dominance of 17–18, 27, 104
  extreme heat exposure 27, 115
  front-of-mind visuals 107
  Global South 17
  The Guardian 120
  heatwave trope 13
  'the idea of heat' 12–14, *14*
  standalone pictures 105
  vulnerable people 15
  see also heatwave visuals

G

Gabbatiss, Josh 50–51
gatekeeping 89, 108
GB News 78
gendered iconography 41
gendered images
  climate protests 94
  flooding 43–44
  war and famine 2
generative-AI (Gen-AI) 2, 9, 121–122
generic images 49, 105–106, 108
Germany 81
'getting on with it' visual trope 42–43
Getty Images
  alternative heatwave imagery 16, *16*
  climate reporting 49, 106
  rising sea levels in Tuvalu 20, *20*
  'smokestacks' searches 49–51
  'wind turbine' image searches 52, 55, 109–110
Gillard, Julia 85–86, *86*
Girifushi, Maldives 90–91, *90*
Gleick, P.H. 35
global climate modelling 18
globalised media ecosystem 22

159

'globally warm' scarf (Sheldon) 67
Goodwillie, Andrew 72, 74, 82
Google Images 111
Gore, Albert (Al Gore) 47
Grabe, M.E. 85
Green Climate Fund 26–27
greenhouse gases 6
'Greenwash!' (*Daily Mail*) 52
'Grief March' protest (XR) 92
The Guardian
  Climate Pledge 2019 117
  COP26 (Glasgow) 24–25
  coverage of scientists 84
  Grid (image management system) 110, 121
  heatwave coverage 120–121
  Lismore floods 41, 103
  Miriam Bulivono image 23–24, 109
  Polar bear imagery 32
  social media platforms 114
Guardian Australia 114

H

Hall, Stuart 9, 100
Hansen, A. 106
Harold, J. 80
Hartley-Brewer, Julia (JHB) 79
Hart, Mary Flora 59–60
Hart, P.S. 38
Hart, S. 56
Hau'ofa, E. 18, 19
Hawkins, Ed 66–69, 81, 112
Hayes, Sylvia 56, 88, 89, 106
Hay Festival 68
heat pump images 57, *57*
heatwaves 11–18, 119
  cultural construction 17
  defining 17, 103
  health risks 11–12
  news articles 13
  weather maps 77–79, *78*
heatwave visuals 12–18, 103
  'angry summer' 75–77, *76*
  generic imagery 106
  'idea of heat' 13–14, *14*, 106–107
  marginalising vulnerable people 15
  portraying risk of heat 120–121
  as a positive event 12–13
  see also extreme heat; 'fun in the sun' heatwave imagery
helplessness
  and climate change 109, 113–114
  racial stereotyping 44–45
  smokestacks imagery 12
Highwood, Ellie 67–69, *68*
'Home Control Room' 60
Hossain Chowdhury, Zakir 42
Hossain, Muhammad Amdad 43
humanitarian aid imagery 22
Hurricane Katrina 44, 45
Hybrid Media System (Chadwick) 101

I

ice imagery 30–31, *31*
'icons' of climate change 122–123
ICUN (International Union for Conservation of Nature) Red List 29
'idea of heat' visual frame 13–14, *14*, 106–107, 120
identifiable people 84
image aesthetics 115
image agencies 105–107
  power 106, 108, 117
  shaping visual discourses 108, 117
  visual meaning-making 49, 52
image copyright data 106
image libraries 3, 22, 119–120
'Image macros' 97
image management systems 110, 121
  see also content management systems
images
  actively read and (re)interpreted 3
  'affective imagery' 107–108
  compositional elements 53–54
  globalised media ecosystem 101
  interrogation 101
  labelling and tagging 111, 121
  movement across interfaces and networks 100–101
  multiple reuses 20, 24, 108–109
  polysemic nature 52, 60–61
image selection 103, 104–105, 110
image-text congruence 14, 114–115
'imaginaries' and 'imagination' 3
implicit visual cues 43–45
An Inconvenient Truth (film) 32, 47
The Independent 50–51
India Times 50
individual frames 5
industrial pollution 47, 48
information architecture 101, 108–116
Instagram 36, 115
Insulate Britain 60
intergenerational equity 94
The Invisible Circus 91
IPCC
  Third Assessment Report 63, 66
  Fifth Assessment Report 66, 81
  Sixth Assessment Report (AR6) 66
  AR6 Synthesis Report *64*, 71, 73
  adaptation 11
  'Burning Embers' diagram 62, 63–66, *64*, 115–116
  Climate Generations figure *71*, 72–73, *74*, 80
  climate impacts 29
  sea levels and small islands 18
  SREX report 38

INDEX

island imagery 19–22, *20*, 24
  see also sea level rise
island 'smallness' 18–19, 27
iStock 22
'It's called summer' meme 77, *78*

J

Jarillo, S. 19, 25
Jaspal, R. 38, 42–43
Johnson, Boris 88–89, *88*
Johnstone, Alastair 17, 105
Jones, Phil 33
journalists 103, 117, 121

K

Kahle, S. 44, 45
Karlsson, M. 113
Kennedy, F. 92
Kitara, Taukiei 18, 26, *28*
Knut (orphaned polar bear cub) 32
Knutti, R. 72
Kofe, Simon 25
Krause, Amber 60
Krebs, H.B. 45
Kyoto Agreement 86

L

land-ocean relationships 18–21
land reclamation projects (TCAP) 27, *28*
landscape photography 54
Lange, Dorothea 41
Leat, Alex *59*
Leeuwen, T. van 95
Lette, Kathie 21
Leviston, Zoe 89–90
Lewis, James 27
Lewis, Rose 58, *59*
LGBTQ+ Pride colours 70
Libarkin, J.C. 81–82
licensing images 22, 24
limitations 10
Lismore, New South Wales 40–41, *40*, 103
London 60
Lorenz, S. 81
Lück, J. 87, 88

M

Machen, R. 3
Machin, D. 106
Mahony, Martin 63–65, 115–116
Maldives 90–91, *90*
Malla, Ranadheer (Ranu) 16, 49, 52, 120
man lying dead in Wuhan image 114
manual labourers 16, *16*
mass protests *35*, 93–95
Matacawalevu, Fiji 109
Matt cartoons (*The Telegraph*) 33, *34*
Mauna Loa Observatory, Hawaii 122–123

Maurice Smith, David 39–41, *40*, 103, 118
McGivern, Aidan 77, *78*
McMahon, R. 72
Me 3645 Studio 57–58
Meares, Andrews *86*
media framing 4–5
media organisations 117
media visual coverage
  extreme heat 12–13
  politicians 93
  protesters 89–90, 93–94
Melbourne 34–35, *35*, 93
memes 73, 77, *78*, 95–98, *96*, 114–115
  see also social media
metadata 110
#MetsUnite 70
Michalec, Ola 58–59
migrant mother iconography 41–42
'migrant mother' image (Lange) 41
migrants crossing border image
  (Mitchell) 106
Mildenberger, M. 82
Miriam Bulivono image 21–25, *21*, 107, 109
  see also flooding images
misinformation 79
Mitchell, Jeff 106
Mittermeier, Cristina 36
Le Monde 73
Mookherje, Nayanika 9
Morpeth floods (September 2008) *39*
'The mother who saved her child from the
  rainstorm' (Hossain Chowdhury) 42
Moutinho, Vera 117–119
movement of images 100–101
multimodal communicative forms 95

N

Nakate, Vanessa 94, 103–104
NASA 74
Nasheed, Mohammed 90–91, *90*
National Geographic 30, 31–32, 36–37
National Museum of Australia 75, 113
National Severe Weather Warnings risk
  matrix (Met Office) 65
NBC 73
Nerlich, B. 38, 42–43
Netflix 70
New Orleans City 45
News Corporation 84
news desks (beats) 84, 103
news organisations 24, 102–105
newspaper ownership 84
newsrooms
  cost-management 24
  'fun in the sun' heatwave imagery 17
  image selection 104
  impacting visual discourses 108
  rights-managed photography 119

thumbnail pictures 112–113
visual climate communication strategies 121
news values 14, 42, 102–103
non-violent civil disobedience 92–93
Norbert, Rosing 31–32
Northern Hemisphere summer solstice 70
'Northern Lights' advertisement (Coca-Cola) 30
Northern Rivers region, New South Wales 39–40, *40*
Norwegian media 52

O

Ocasio-Cortez, Alexandria 98
ocean-land images 19–21, *20*
ocean states 18
Oliver, Neil 78
online courses 117
online media 20, 115
online opinion shapers 95–98
Open Planet 119
Oxford Climate Journalism Network (OCJN) 117

P

Pacific Climate Warriors 24–25, 26–27
Pacific islanders 18, 25–27
Pacific island imagery 109
Pakistan 44
Paris Agreement 60
Pearce, Warren 111, 115
Pearson, Allison 79
Peeters, Theo 15
performance protests 90–92, *90*
personalisation 84
person-centred climate-energy imagery 56–58
Philadelphia Enquirer 47
photo editors 108
photojournalism 106
photos *see* images
platform formats (aesthetics) 114–115
Plumb, G. 80–81
polar bears
  apex predators 30, 31
  cartoons 32–34, *34*
  costumes 34–35, *35*, 93
  friendly, cuddly and huggable 32
  icon of climate change 35
  imagery 29–37, 31, 36, 107, 115
  popular culture 36–37
  population change 30
  'threatened' species 29
political imagery 84–89, *86*, *88*
politics desks 84–85, 103
Price, V. 5
Pritchett, Matt 33, *34*
protests 89–95
  mass protests *35*, 93–95
  performance protests 90–92, *90*
public debates 60
public health messages 15
Público (newspaper) 119
purple colour controversy 75–77, *76*, 113

Q

Q-method study 37–38, 89, 94

R

racialised images of war and famine 2
racial stereotyping 44
racism 2
racist visual narratives 106
Radtke, Alexander 72
Reading Football Club 70
'Reasons for concern' (IPCC) *see* 'Burning Embers' diagram (IPCC)
Reddit 115
Red Rebel Brigade 91–93
reusing images 20, 24, 108–109
Reuters 91, 106
Reuters Digital News 118
reverse image searches 20, 50–51, 111
Ribbans, Elisabeth 114
Rich, Ben 17–18
rights-managed photography 119
Rio Olympics 2016 67
rising sea levels *see* sea level rise
risks of heat images 120
Rivers, Damian 96, *96*
Roberts, Helen 65
rock art 3
Roe, Stephanie 74
Rose, Gillian 3, 100
Rosing, Norbert 30
Ross, Andrew S. 96, *96*
Royal Meteorological Society 80–81
Ruane, Alex 72
Ruetenik, G. 81–82
Russell, Stuart 2

S

'sceptic' logic memes 96–98, *96*, 114–115
Science (journal) 35
science literacy 82
scientific knowledge 18–19
scientific visuals
  climate projections 81–82
  going viral 112, 113
  as 'showplace of science' 62–63
scientists 1, 84
SeaLegacy 36
sea level rise 18–27
  challenging visual narratives 25–27
  coral atolls 19–21, *20*

imagery and agency of islanders 24–25
   visual tropes 21–24, *21*, 25
'sea of islands' (Hau'ofa) 18
Seeneen, Mohammed 90
self-efficacy 56
sexism 2, 44
Sharma, Alok 89
Sheldon, Joan 67
Shields, F. 117, 120
Shifman, L 95
#ShowYourStripes 70, 112
Shutterstock, Inc. 111
Silva, Marco 116
Simonsen, Anne Hege 52
'sinking islands' visual trope 25
'Six Americas' 81
'Skolstrejk för klimatet' (School Strike for Climate) 93
small islands 18–21, 24, 27
smokestacks imagery 7, 47–51, *48*, 107, 113–114
social media 73, 112, 113
   see also memes
social sciences 4, 82
socio-cultural barriers 44
solutions journalism 15, 118
Solutions Journalism Network (SJN) 58
solutions visuals 119
Sontag, Susan 9
Spiegel, S.J. 109
'sponsored posts' 97
Squire, Justine 91
Stackhouse, Shawndu 16, *16*
standalone newspaper pictures (standalones) 105
starbursts 106
starvation 36–37
Stauffacher, M. 72
stereotyped images 44–45
stock images 22, 49, 106
stock libraries *see* image libraries
Strauss, David Levi 9
Studio Silverback 119
Sturgeon, Nicola 103
The Sun 84
surveys 38
synecdoches *34*, 51, 108

T

tagging images 111, 121
Takahashi, B. 82
talking heads photographic style 88–89, *88*
TalkTV 78–79
Tama, Mario 19–21, *20*, 25
The Telegraph 30, 32, 33, *34*, 79
'temperature blankets' 67–68, *68*
Teriete, C. 57

Terminator robots 2
Tewksbury, D. 5
text plus image memes 97
thermometers 14, *14*, 103, 106
Thomas, S.R. 81–82
Thomson, T.J. 104
Three Mile Island nuclear accident 47
thumbnail pictures 112–113
Thunberg, Greta 70, 93, 94, 103
The Times 24, 42, 51, 55
The Times of India 50
TinEye reverse image searches 20, 50–51
Tony Auth cartoons 47
training programmes 117–118, 121
Transition Belsize 57
Trump, Donald 95
Turning Point USA 98
Tuvalu 19–21, *20*, 23, *23*, 25
Tuvalu Coastal Adaptation Project (TCAP) 26–27, *27*, *28*
Twitter *see* X
two-way slider images 15

U

UBC (Uganda Broadcasting Corporation) 70
Ultra Low Emissions Zones (ULEZ) 60
United Kingdom (UK)
   climate communication 10
   climate-energy case studies 58–59
   electricity pylons 46
   greenhouse gas emissions targets 50
   Health Security Agency 12
   House of Lords 58
   Meteorological Office (Met Office) 65, 77, *78*, 79
   newspaper visual coverage of climate change 84, 87
   Office for National Statistics 12
   visual communication studies 38
   weather forecast colour hues 77
   weather map meme 77–78, *78*, 116
   'weather talk' 12
United Nations Framework Convention on Climate Change (UNFCCC) 85
   Article 2 63
   Green Climate Fund 26–27
   see also Conference of the Parties (COP)
United Nations (UN) 43
United States (US)
   Endangered Species Act (ESA) 32
   engagement with climate change 81
   'fun in the sun' visual frame 17
   Great Depression 41
   Green New deal 98
   House Select Committee on the Climate Crisis 69
   newspaper visual coverage of climate change 84

visual communication studies 38
visual representations of floods 43–44
University of East Anglia 32–33, 86, 122

V

van Leeuwen, T. 95
Vevea, N.N. 45
viral images 4
  memes 77, 78, 96–97, 96
  polar bear imagery 36
  scientific visuals 8, 112, 113
  see also Climate Spiral (Hawkins, Fæhn and Fuglestvedt); Climate Stripes
visual aesthetics 26, 44, 115
visual framing 5, 13–14, 14
visual gatekeeping 89, 108
visual imaginings 2
visualising rising sea levels 21, 21
visual libraries *see* image libraries
visual metonyms 107–108
visual news coverage 45, 87
visual news ethnography 88
visual tropes 6, 25, 108–109
Vlasceanu, M. 113

W

Walczak, Janusz 48
Walk Against Warming protest (Melbourne, 2009) 34–35, 35, 93
'wallpaper of consumer culture' (Frosh) 105
Warming Stripes 68–73, 68, 71, 74, 80–81
'warrior aesthetic' (Farbotko and Kitara) 26

Washington Post 73
weather maps 74–80, 76, 78, 113
Weaver, Iain 97
Webster, R. 57
Wessler, H. 87
Western narratives of island vulnerability 18–19, 20–21, 20, 25–26, 27
Westwood, Ned 49, 50, 52, 97–98
Whiteside, E. 44, 45
Whitmore, Greg 105
wide angle lenses 55
Willy Wonka & the Chocolate Factory film 96
wind turbine images 7, 51–55, 53, 60, 110
'wishful sinking' narratives 21, 23, 23
World Health Organisation (WHO) 12
Worth, Kiara 88, 88
Wozniak, Antal 87, 95

X

X (Twitter) 43, 112, 115

Y

Yale Environment 360 website 13
Yale Program on Climate Change Communication 81
Young, N. 14, 115
young protesters 94
Yu, N. 44, 45

Z

Zwickle, A. 82

www.ingramcontent.com/pod-product-compliance
Lightning Source LLC
Chambersburg PA
CBHW071708020426
42333CB00017B/2194